Land and Hydropolitics in the Nile River Basin

The Nile River Basin supports the livelihoods of millions of people in Egypt, Ethiopia, Sudan, South Sudan and Uganda, principally as water for agriculture and hydropower. The resource is the focus of much contested development, not only between upstream and downstream neighbours, but also from countries outside the region. This book investigates the water, land and energy nexus in the Nile Basin.

It explains how the current surge in land and energy investments, both by foreign actors as well as domestic investors, affects already strained trans-boundary relations in the region and how investments are intertwined within wider contexts of Nile Basin history, politics and economy. Overall, the book presents a range of perspectives, drawing on political science, international relations theory, sociology, history and political ecology.

Emil Sandström is Senior Researcher at the Nordic Africa Institute, Uppsala, Sweden, and Senior Lecturer and former Head of the Rural Development division at the Swedish University of Agricultural Sciences (SLU).

Anders Jägerskog is Counsellor for regional water issues in the Middle East and North Africa for the Embassy of Sweden in Amman, Jordan, and Associate Professor at the School of Global Studies at Göteborg University, Sweden. He is former Head of the Transboundary Unit at the Stockholm International Water Institute (SIWI).

Terje Oestigaard is Senior Researcher at the Nordic Africa Institute, Uppsala, Sweden, and Docent in Archeology at Uppsala University.

Earthscan Studies in Water Resource Management

Water, Power and Identity
The cultural politics of water in the Andes
Rutgerd Boelens

Water and Cities in Latin America
Challenges for sustainable development
Edited by Ismael Aguilar-Barajas, Jürgen Mahlknecht, Jonathan Kaledin and Marianne Kjellén

Catchment and River Basin Management
Integrating science and governance
Edited by Laurence Smith, Keith Porter, Kevin M. Hiscock. Mary Jane Porter and David Benson

Transboundary Water Management and the Climate Change Debate
Anton Earle, Ana Elisa Cascão, Stina Hansson, Anders Jägerskog, Ashok Swain and Joakim Öjendal

Rules, Norms and NGO Advocacy Strategies
Hydropower development on the Mekong River
Yumiko Yasuda

The Mekong: A Socio-legal Approach to River Basin Development
Ben Boer, Philip Hirsch, Fleur Johns, Ben Saul and Natalia Scurrah

Trans-jurisdictional Water Law and Governance
Edited by Janice Gray, Cameron Holley and Rosemary Rayfuse

Water and Rural Communities
Local politics, meaning and place
Lia Bryant with Jodie George

Land and Hydropolitics in the Nile River Basin
Challenges and new investments
Edited by Emil Sandström, Anders Jägerskog and Terje Oestigaard

For more information and to view forthcoming titles in this series, please visit the Routledge website: http://www.routledge.com/books/series/ECWRM

Land and Hydropolitics in the Nile River Basin

Challenges and new investments

Edited by Emil Sandström, Anders Jägerskog and Terje Oestigaard

Routledge
Taylor & Francis Group
LONDON AND NEW YORK

First published 2017 by Routledge

2 Park Square, Milton Park, Abingdon, Oxfordshire OX14 4RN
711 Third Avenue, New York, NY 10017

Routledge is an imprint of the Taylor & Francis Group, an informa business

First issued in paperback 2018

British Library Cataloguing-in-Publication Data
A catalogue record for this book is available from the British Library

Library of Congress Cataloging in Publication Data
Names: Sandstroem, Emil, editor. | Jaegerskog, Anders, editor. | Oestigaard, Terje.Title: Land and hydropolitics in the Nile River basin : challenges and new investments / edited by Emil Sandstroem, Anders Jaegerskog and Terje Oestigaard. Description: London ; New York : Routledge, 2016. | Series: Earthscan studies in water resource management | Includes bibliographical references and index.Identifiers: LCCN 2016009964| ISBN 9781138921757 (hbk) | ISBN 9781315686172 (ebk)Subjects: LCSH: Water-supply--Nile River Watershed--Management. | Water-supply--Political aspects--Nile River Watershed. | Water security--Political aspects--Nile River Watershed. | Food security--Nile River Watershed.Classification: LCC HD1699.N5 L36 2016 | DDC 333.730962--dc23LC record available at https://lccn.loc.gov/2016009964

ISBN: 978-1-138-92175-7 (hbk)
ISBN: 978-0-367-02969-2 (pbk)

Typeset in Bembo
by GreenGate Publishing Services, Tonbridge, Kent

Contents

Illustrations

Figures

Tables

Acknowledgements

This book has its origin in the research project entitled Water Politics in the Nile Basin – Emerging Land Acquisitions and the Hydro-Political Landscape financed by the Swedish Research Council and implemented by the Nordic Africa Institute, the Stockholm International Water Institute (SIWI) and the Swedish University of Agricultural Sciences (SLU), together with various partner institutions in the Nile Basin. We greatly acknowledge the support provided by the Swedish Research Council and our host institutions.

Some of the results have been disseminated during a number of conferences and workshops, and the editors acknowledge the insights and valuable comments during these events. Three events deserve particular gratitude. First, the session Investments in Land & Water in the Nile Basin and TWM Nexus held at the 2013 World Water Week, Stockholm. Second, the Authority, Territory, Community: Regional Dynamics of Water, Land Tenure Regimes and State Formation in Africa workshop, organized in Kampala in October 2013. Finally, the workshop Transformation in Smallholder Agriculture in and Around the Lake Tana Basin held in Bahir Dar in Ethiopia in August 2015. With regards to these workshops, we would like to thank SIWI, the Nile Basin Programme, Professor Edward K. Kirumira, Dr Tore Saetersdal, Dr Anders Sjögren, Dr Opira Otto, Dr Gedef Abawa Firew and the Vice President of Research and Community Service of the Bahir Dar University, Dr Tesfaye Shiferaw.

Finally, as editors we would also like to express our gratitude to all contributors for their outstanding cooperation throughout the project and, in particular, during the final months. A sincere appreciation to Ashley Wright and Tim Hardwick from Routledge for their invaluable guidance in the process of making this book, and the two anonymous reviewers.

Uppsala, Sweden and Amman, Jordan, May 2016.
Emil Sandström, Anders Jägerskog and Terje Oestigaard

Contributors

Editors

Emil Sandström is Senior Researcher at the Nordic Africa Institute in Uppsala, Sweden, where he conducts research on Nile Basin-related issues. He is also Senior Lecturer and former Head of the rural development division at the Swedish University of Agricultural Sciences (SLU) in Uppsala. His research and teaching interests are interdisciplinary in character and embrace perspectives found in political ecology and rural sociology in Eastern Africa and Sweden. Besides teaching and research, he has worked with research-based assignments and commissions in different capacities and for different organisations – for example, the Swedish Ministry for Foreign Affairs, the International Fund for Agricultural Development and the Swedish International Development Cooperation Agency.

Anders Jägerskog is Counsellor for regional water issues in the Middle East and North Africa for the Embassy of Sweden in Amman, Jordan. He is also Associate Professor at the School of Global Studies at Göteborg University as well as former Head of the Transboundary Unit at the Stockholm International Water Institute (SIWI). He has worked at the secretariat for the Expert Group on Development Issues at the Swedish Ministry for Foreign Affairs, as Senior Programme Manager, water resources, at the Embassy of Sweden, Nairobi, as well as at the Stockholm International Peace Research Institute (SIPRI) with Middle Eastern security issues. The views expressed by him do not necessarily reflect the views of the Swedish International Development Cooperation Agency or the Swedish government.

Terje Oestigaard is Senior Researcher at the Nordic Africa Institute, Uppsala, and docent in archaeology at Uppsala University. His recent books include: T. Tvedt and T. Oestigaard (2016) *A History of Water, Series 3, Vol 3, Water and Food: From Hunter-Gatherers to Global Production in Africa*, I.B. Tauris, London; K. Havnevik, T. Oestigaard, T. Virtanen and E. Tobisson (2016) *Framing African Development: Challenging Concepts*, Brill, Leiden; and T. Oestigaard (2015) 'Dammed Divinities: The Water Powers at Bujagali Falls, Uganda', *Current African Issues*, 62, Nordic Africa Institute, Uppsala.

Contributors

Atakilte Beyene is Researcher at the Nordic Africa Institute. He holds a PhD in Rural Development Studies. He has been working in universities and research institutes in Ethiopia and Sweden. His research focuses on agrarian and rural institutions, natural resources management, food security and gender studies. He has conducted extensive field studies in Ethiopia and Tanzania. He has both coordinated and worked on many interdisciplinary research projects and networks in the Nordic and African countries. His current research includes studies on large-scale agricultural investments, irrigation management and smallholder agriculture in Africa.

Ana Elisa Cascão is Programme Manager at the Transboundary Water Unit at SIWI, Sweden. She is a political scientist with a PhD in the political economy of water resources management in the Nile Basin (in King's College London, UK). For the past six years at SIWI, she has been participating in numerous academic projects and consultancy assignments in collaboration with regional and international institutions, such as the Nile Basin Initiative (NBI), the Eastern Nile Technical Regional Office (ENTRO), the International Water Management Institute – East Africa, the United Nations Development Programme Water Governance Facility, the Nile Basin Trust Fund and the World Bank Cooperation in International Waters in Africa.

Ramy Lotfy Hanna is a doctoral researcher at the Institute of Development Studies, University of Sussex in the UK. His research interests encompass resources politics, sustainable use of natural (land/water) resources and transboundary water governance. His academic background is economics and international relations, and he earned his Master's degree in Environmental Studies from York University in Toronto. His professional portfolio as an economic development researcher and an environment specialist in Egypt and the Middle East and North Africa (MENA) region includes a number of agencies such as the United Nations Economic Commission on Africa, the African Development Bank, the Organisation for Economic Co-operation and Development, the Canadian International Development Agency (CIDA)/Global Affairs Canada and the United Nations Environment Programme, among others. As an environment specialist, he also conducted the evaluation of the Nile Basin Discourse and advised the Eastern Nile Watershed Management Project.

Mats Hårsmar is currently *chargé d'affaires* at the Swedish Embassy in Ouagadougou, Burkina Faso. Previously he served as Senior Researcher and Research Leader at the Nordic Africa Institute in Uppsala, Sweden. He has also held positions as Chief Analyst for development issues and Head Secretary for the Expert Group on Development Issues secretariat at the Swedish Ministry for Foreign Affairs, as well as Programme Officer at the Swedish Expert Group for Aid Studies. He holds a PhD from the SLU.

Martin Keulertz is a visiting Assistant Professor at the American University of Beirut (Lebanon), visiting research fellow at Texas A&M University (USA) and a research fellow at King's College London (UK). Prior to this, he worked as a postdoctoral fellow in the Department for Agricultural and Biological Engineering at Purdue University, US, and as a research fellow in the Department for Agricultural Economics at Humboldt University Berlin. In 2013, he obtained his PhD at King's College London, UK. His research interests centre on water-smart agricultural subsidies, responsible investment and water in food supply chains.

Kyungmee Kim is a PhD candidate at the Department of Peace and Conflict Research at Uppsala University in Sweden, and a member of the research school under the United Nations Educational, Scientific and Cultural Organization (UNESCO) centre for international water cooperation. Her research interests include water and natural resources, conflict and security. In her dissertation project, she focuses on the links between armed conflict and water and natural resources in Myanmar. Prior to starting her PhD study, she worked for three years on transboundary water management at SIWI.

Alan Nicol is a political scientist with over 20 years of experience on the Nile Basin. He works at the interface between research, policy and development practice, which includes work with government and intergovernmental institutions, multilateral and bilateral agencies, research institutes, global think tanks and international non-governmental organizations. His key concern is multilevel analysis of water policy and practice, and the relationship between different 'waters' (domestic, agricultural, transboundary and global) and other areas – including climate change, the environment, energy, economic growth and social change. He has particular knowledge of governance, political economy and sustainable livelihoods analysis, and has led field research, policy analysis, teaching, training and consulting activities in sub-Saharan Africa, the Middle East and North Africa, and South and South-East Asia. He is currently a principal researcher and theme leader on governance, gender and poverty at the International Water Management Institute.

David Ross Olanya, Gulu University, Uganda, and University of Antwerp, Belgium, specializes in water and land issues in Uganda. His research interests also include climate change and adaptation regimes, water politics and governance, environmental security, and land and agrarian transformation. He has been a World Social Science Fellow 2013, Visiting Research Fellow 2011 at African Nordic Institute (Uppsala), and Mo Ibrahim Foundation Fellow 2011, School of Oriental and African Studies, University of London.

Harry Verhoeven is Assistant Professor of government at the School of Foreign Service of Georgetown University, based in Qatar. He is the convenor of the Oxford University China–Africa Network and the author of *Water, Civilisation and Power in Sudan: The Political Economy of Military-Islamist State Building* (Cambridge University Press, 2015). His forthcoming book, co-authored with Philip Roessler, is entitled *Why Comrades go to War: Liberation Politics and the Outbreak of Africa's Deadliest Conflict.* (Hurst/Oxford University Press, 2016).

Wondwosen Michago Seide is a PhD student at Lund University. He was a Louse Water Scholar at Oxford University where he received his MSc in Water Science, Policy and Management. Over the past ten years, he has worked as a researcher and consultant with regional and international experience in the area of water, land, climate change, environment and development for various regional and international organizations, such as the International Institute for Environment and Development; the United Nations Development Programme; the Institute of Development Studies, University of Sussex; the Institute for Security Studies; the Ethiopian International Institute for Peace and Development; the Intergovernmental Authority on Development; and the Nile Basin Discourse Forum.

Abbreviations

AAAID	Arab Authority for Agricultural Investment and Development
ADFD	Abu Dhabi Fund for Development
AFESD	Arab Fund for Economic and Social Development
AOAD	Arab Organization for Agricultural Development
bcm	billion cubic metres
BIT	Bilateral Investment Treaty
BRIC	Brazil, Russia, India and China
CAADP	Comprehensive Africa Agriculture Development Program
CFA	Cooperative Framework Agreement
CIRAD	Centre de Coopération Internationale en Recherche Agronomique pour le Développement
COMESA	Common Market for Eastern and Southern Africa
COP	Conference of the Parties
EIA	Environmental Impact Assessment
ENSAP	Eastern Nile Subsidiary Action Programme
ENTRO	Eastern Nile Technical Regional Office
FAO	Food and Agriculture Organization
FDI	foreign direct investment
FDRE	Federal Democratic Republic of Ethiopia
GCC	Gulf Cooperation Council
GDLA	Global Dry Land Alliance
GDP	gross domestic product
GERD	Grand Ethiopian Renaissance Dam
GIGA	German Institute for Global and Area Studies
GIZ	Deutsche Gesellschaft für Internationale Zusammenarbeit
GRS	Gambella Regional State
GTP	Growth and Transformation Plan
HI	Al-Harakat Al-Islamiyyah
ILC	International Land Coalition
IPoE	International Panel of Experts
IWRM	integrated water resource management
JMP	Joint Multipurpose Project

KAISAIA	King Abdullah Initiative for Saudi Agricultural Investment Abroad
KDCIS	Koga Dam and Canal Irrigation Scheme
KSC	Kenana Sugar Company
LAS	League of Arab States
LSI	large-scale irrigation
LSLAs	large-scale land acquisitions
masl	metres above sea level
mcm	million cubic metres
MENA	Middle East and North Africa
MPIDP	Megech Pump Irrigation and Drainage Scheme
MT	metric tonnes
MW	megawatts
MWh	megawatt hours
NBI	Nile Basin Initiative
NWA	Nile Waters Agreement
ODA	official development assistance
OPEC	Organization of the Petroleum Exporting Countries
QIA	Qatar Investment Authority
QNFSP	Qatar National Food Security Programme
SIWI	Stockholm International Water Institute
SLU	Swedish University of Agricultural Sciences
SWF	sovereign wealth fund
TECCONILE	Technical Cooperation Committee for the Promotion of the Development and Environmental Protection of the Nile
TWM	transboundary water resources management
UAE	United Arab Emirates
UNESCO	United Nations Educational, Scientific and Cultural Organization
WNSC	White Nile Sugar Company

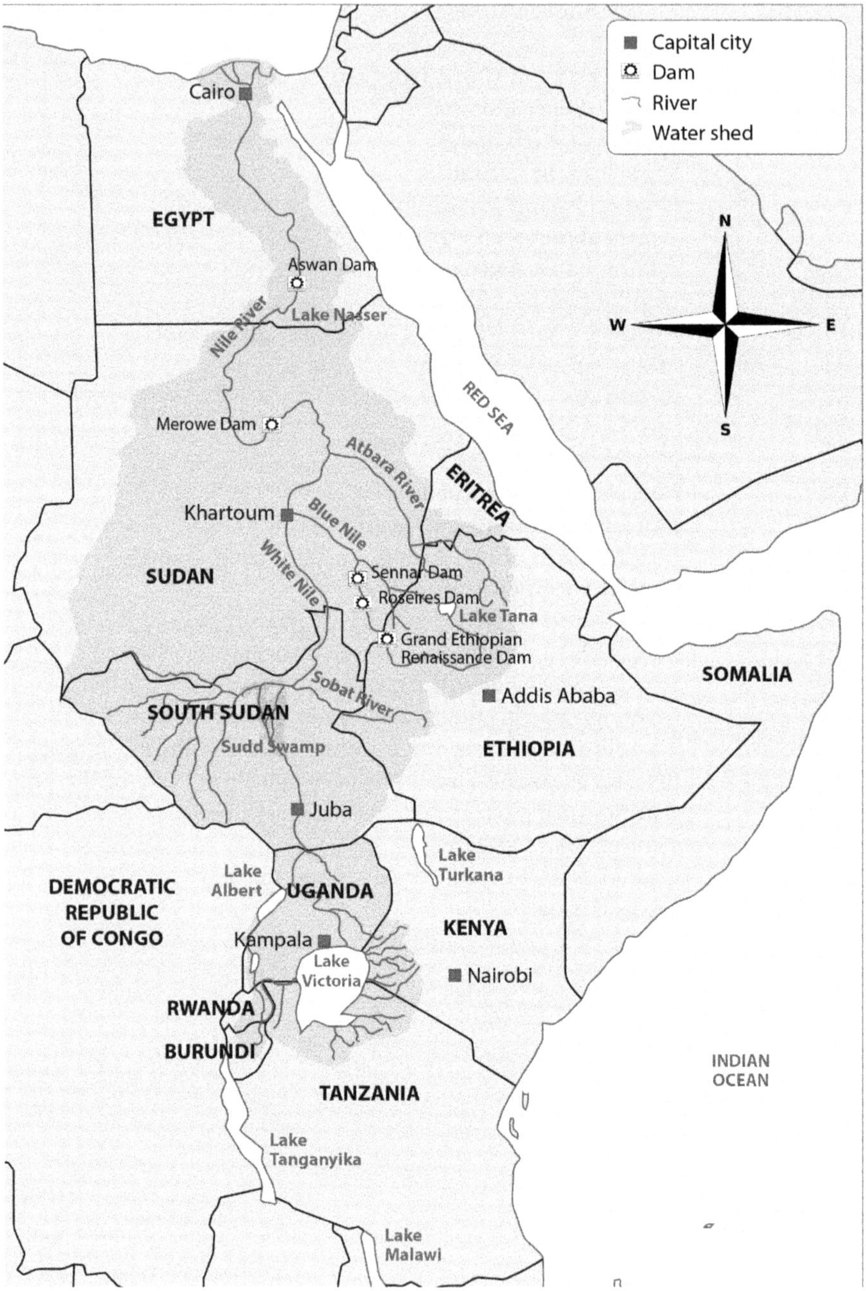

Figure 0.0 Map displaying the Nile Basin and the riparian states

Design: Henrik Alfredsson, adapted from the Nile Basin Research Programme 2009.

1 Changing challenges

New hydropolitical landscapes in the Nile Basin

Emil Sandström, Anders Jägerskog[1] *and Terje Oestigaard*

Buy land, they're not making it anymore.
(Mark Twain)

The political hydro-geography of the Nile

Land and hydropolitics in the Nile River Basin are intimately interwoven and new water-related investments in agriculture and hydropower pose increasing challenges and changing hydropolitical landscapes. Not only are the physical landscapes of the Nile changing, but so are the political landscapes. Studies of Nile Basin hydropolitics have been a reoccurring theme in academic studies over the years (see, e.g., Cascão, 2012; Tvedt, 2004, 2010a; Waterbury, 2002; Whittington *et al.*, 2005). The relations between the riparian countries in the basin are complex given the hydrological nature of the basin, varying hydroclimates in different parts of the region and high variation in levels of development in the region.

This book analyses yet another layer that further contributes to the hydropolitical complexity of the basin – namely, the current surge in land and energy investments. There have been widely differing estimates of the scale of which farmland and water-related investment are being acquired, but suffice to say that the Nile Basin has not seen cross-border land acquisitions and water-related investments on this scale before. Parallel to these events a number of countries in the Nile Basin have also undergone several profound political changes, with transboundary implications. With this background, the contributors in this book analyse the changing transboundary water relations in the region in the context of increasing investments in energy and agriculture and discuss this from a range of perspectives.[2] The various chapters address how the recent surge of land- and water-related investments affect the transboundary water relations in one of the largest and most challenging river basins in the world, where contemporary developments, in particular in the Eastern Nile region, have played a pivotal role for reshaping the hydropolitical landscapes of the Nile Basin. Together, the chapters in this book provide new insights into a number of key aspects and events relating to both past and present land and energy investments and discuss how they are intertwined in different hydropolitical and development trajectories in the Nile Basin.

The Nile River is regarded as the longest river in the world and the water catchment area (basin) comprises about 10 per cent of the total land area of Africa (about 3,173,000 square kilometres). The Nile Basin is shared by 11 states – Burundi, the Democratic Republic of Congo, Egypt, Eritrea, Ethiopia, Kenya, Rwanda, Tanzania, South Sudan, Sudan and Uganda – that together inhabit about 437 million people, out of which 238 million live within the basin (NBI, 2013). The population in the 11 countries is expected to grow rapidly and reach about 591 million in 2025 and 861 million by 2050 according to UN projections (UN, 2009) of which approximately 54 per cent will live in the basin itself being dependent upon the Nile waters. The population growth is expected to put increasing pressure on water resources for food and development in all basin countries. According to Salman (2011), who draws on Waterbury (2002), the stakes and interests of Egypt, Sudan, South Sudan and Ethiopia are classified as very high; those of Uganda as high; those of Burundi, Rwanda, Kenya and Tanzania as moderate; and those of the Democratic Republic of Congo and Eritrea as low.

The Nile is the lifeline of Egypt, but also crucial to the development of Ethiopia, Sudan, South Sudan and Uganda, although in different ways. In Egypt, the Nile is not only the main source of prosperity, but also her very existence depends on the Nile. Of its two main sources, the Blue Nile flows from Lake Tana in Ethiopia, while the White Nile flows from Lake Victoria in Uganda. About 85 per cent of the water in the Nile originates from the Ethiopian tributaries while the remainder comes from the White Nile (Waterbury, 2002).

The free flow of the Nile has historically been a national security issue for Egypt and the country has held on to the colonial water agreement from 1929[3] and the later 1959 agreement, which provides Egypt with the lion's share of the water. The 1959 agreement between Sudan and Egypt established the average annual flow of the Nile to 84bcm (billion cubic metres) of which Egypt's share is 55.5bcm, Sudan 18.5bcm and the remaining 10bcm represent the evaporation losses from the Aswan High Dam reservoir at Lake Nasser (Waterbury, 2002). Tanzania, Uganda and Kenya, on the other hand, held on to the Nyerere Doctrine,[4] and they have argued that the agreements from 1929 and 1959 were signed when most of the upstream countries were colonies and therefore those treaties should not be binding. Ethiopia, being independent when the agreements were signed, argues that they were not even consulted and therefore regards the water agreements as non-binding.

In 1999, the Nile Basin Initiative (NBI) was established among the then ten Nile countries (South Sudan became independent on 9 July 2011), with Eritrea having observer status. After a decade, the Cooperative Framework Agreement (CFA) was signed in 2010 by five of the riparian countries (Ethiopia, Tanzania, Uganda, Rwanda and Kenya), with Burundi following suit in 2011. According to the agreement, 'Nile Basin States shall in their respective territories utilize the water resources of the Nile River system and the Nile River Basin in an equitable and reasonable manner'. One major challenge with this agreement is that the whole agrarian and irrigation question is not discussed, which is the largest water consumer. Egypt and Sudan have opposed the agreement.

Egypt's strategy has essentially been to prevent the emergence of any states that could challenge its domination over the Nile water, and to obstruct any development along the banks of the Nile that could either redirect the flow of the water or reduce its volume (Kendie, 1999; Muhammad, 2010). This strategy has historically been secured by military means and threats, but also through influence in international institutions. For example, Egypt has on several occasions been accused of supporting Eritrea and Somalia in the wars against Ethiopia (Kendie, 1999; Muhammad, 2010) and political leaders from Egypt have threatened to go to war if the free flow of the Nile was perceived to be in danger.

In the media the Middle East and, in particular, the Nile Basin have often been portrayed as the most likely hot-spots where 'water-wars' may erupt. As the late President Sadat stated: 'Any action that would endanger the waters of the Blue Nile will be faced with a firm reaction on the part of Egypt, even if that action should lead to war' (cited in Kendie, 1999). Similar statements have been delivered by the former presidents Hosni Mubarak and Mohammad Morsi.

The security of the Nile has been central in Egyptian military planning. There can be no doubt that Egypt has the military muscle to go to war against other Nile Basin states. As an example, Egypt's military expenditure in 2005 was more than twice the equivalent expenditure of all the ten other riparian states together (Adar, 2011: 183). Whether water wars will occur is, however, another question. Water-related conflicts may cause conflicts within states rather than between states and conflict and cooperation may exist simultaneously at different levels at a given time (Zeitoun and Mirumachi, 2008). Thus, Egypt traditionally has been seen as the hegemon in the basin due to its ideational, military and economic domination, despite having the disadvantaged downstream position (Cascão and Zeitoun, 2010).

Analytical and conceptual perspectives

The various chapters have not used a common analytical framework for analysis and there is variation between the chapters, with some being more theoretical and others more empirical and case-study oriented, thus providing different perspectives. The hydropolitical landscapes of the Nile Basin are not confined to the basin's borders and as many of the chapters in this book show, the hydropolitical landscapes of the Nile are in fact comprised by several 'landscapes' that are influenced, as in the past, by actors and factors far away from the Nile Basin itself. Several contributions underscore the complexity and the 'messiness' of hydropolitics, and take their point of departure in understanding the hydropolitics of the River Nile as manifestations of negotiated social practices located both locally and globally and within wider contexts of history, politics and economy. Framing the analysis in this way makes it possible to elucidate how, for instance, various political, economic and legal conflicts intertwine and how relations within and between the riparian states of the Nile River

Basin have developed over time. In such contexts political and cultural symbols of identity, power and nationalism are often brought into play in shaping the outcomes of the resource system.

Several chapters combine both historical and contemporary perspectives and some chapters take the Nile Basin, part of the basin or a country in the basin as a point of departure for the analysis, whereas others put more emphasis on how the Nile Basin hydropolitics are related to developments 'outside' the Nile Basin geography. The book also analyses the land- and water-related investments, both from the perspective of the investor as well as the investee, and highlights the power struggles between and within the riparian states of the River Nile Basin. With regards to the latter aspect, the very terminology used in discourses – by activists, academics, investors and politicians – also reflect and are part of wider geo-political landscapes, with subsequent implications to ongoing land- and water-related processes.

As indicated, Mark Twain once said: 'Buy land, they're not making it anymore', and private individuals, business enterprises and governmental agencies seem to follow this advice and, if impossible to buy, then hiring or leasing is the second best option. The complexity on the ground is, however, challenging simple conceptualizations of these processes. 'Investment' or 'acquisition' is the preferred term for investors, entrepreneurs and even facilitating governments since it alludes to positive development, whereas the terms 'land grab' and 'water grab' allude to the negative consequences. Although the terms are often used interchangeably in the literature, the two notions are still different. Not all land deals are investments in a strict sense and not all agricultural investments involves land deals (Cotula, 2013: 7) and, as Cotula says, 'the polarized debates about the land rush ultimately reflect competing visions about the future of world agriculture, and particularly about the roles of small- and large-scale farming' (Cotula, 2013: 5).

From the perspective of the people affected, '[l]and grabbing (and associated other forms of grabbing) constitutes a recent intensification of an historic threat to rural livelihoods, to democratic governance and equity, and to long-term environmental sustainability' (Edelman *et al.*, 2013: 1529). Elaborating along these lines, Borras and Franco (2013) argue that 'land grabbing' means 'the capturing of control of relatively vast tracts of land and other natural resources through a variety of mechanisms and forms, carried out through extra-economic coercion that involves large-scale capital'.

> Even though the term 'land grabbing' carries a lot of baggage and remains problematic, in our view it still has the advantage of 1) focusing attention on the core issues of politics and power relations; and 2) underscoring the dimension of extra-economic coercion involved in land deals.
>
> (Borras and Franco, 2013: 1725)

This also relates to tenure systems and customary rights, and few places in the world are these institutions and traditions as complex and fundamental as in Africa. On the one hand, the term 'land grab' may in many cases be the most appropriate, since many investments or deals may cause conflicts when people are resettled, land titles are lost and resources and livelihoods changing for the worse. On the other, in many cases the ultimate ownership of land and water is within the state, and strictly speaking a state cannot 'grab' anything it owns in the first place, but it may distribute the resources disproportionally or exclusively.

In this volume, different chapter contributions use different terminologies in their analyses, and no coherency has been sought since, on the one hand, there are various processes involved and described, and, on the other, the terminology depends on perspectives and different emphases by the various actors involved.

Nile geo-politics in a historic perspective

While this book has a contemporary emphasis, today's politics are also trajectories of the past, shaping future policies, and today's shifts in geo-politics can best be understood in a broader historic context where building dams and large-scale agricultural investments also played a crucial role, then as today. Many of the ongoing developments along the Nile were, in fact, first proposed by the British more than 100 years ago but, for economic, historic and political reasons were never implemented. A brief historic analysis may also highlight the contemporary complexity where several of the new water-related investments are positioned. In order to understand the background to the 1929 and 1959 agreements, and consequently the disagreements about the CFA, one has to turn to the colonial era. Terje Tvedt says, because 'in the Nile basin the past is in the present and the present in the past … nobody can escape the impact of the Nile's power and its history' (Tvedt, 2004: 326).

In *The River Nile in the Age of the British: Political Ecology and the Quest for Economic Power* (2004), Tvedt analyses in detail how the British imperial policies in the Nile had one overall aim of securing British interests. Britain took control of Egypt in 1882. Small amounts of cotton produced in the delta were sold into a growing world market from the 1820s but, from the 1860s cotton exports made up about 80 per cent of Egypt's exports. British industry had a great economic interest in Egypt. The Lancashire textile industry aimed to reduce its dependence on American cotton and increase its supply of cheaper Egyptian cotton. To increase productivity and, hence, profits, more Nile water was needed at the right time (the summer season in Egypt) prior to the annual Blue Nile floods, and therefore cotton production was dependent on the waters of the White Nile. The overriding question was how to secure enough water for cotton production and at the same time control potentially devastating floods (Tvedt, 2004: 20–22). Thus, by the end of the nineteenth century, 'increased and improved water control was destined to top the agenda of any administration in Egypt' (Tvedt, 2004: 22). The British were the first who truly saw and tried to exploit this connection and the colonialization in this part of Africa; present-day hydropolitics cannot be understood properly without a water perspective emphasizing the Nile as a river system.

The then Egyptian Prime Minister Nubar Pasha (1884–1888 and 1894–1895) summarized Egypt's situation and position in a famous one-liner: 'The Egyptian question is the irrigation question' (op. cit. Tvedt, 2011: 176), which captures the same essence and importance of the Nile as Herodotus said when he proclaimed that 'Egypt is the gift of the Nile'. The later Lord Lugar, who became the British ruler of Uganda, wrote in the early 1890s: 'Egypt is indebted for her summer supply of water to the Victoria lake, and a dam built across the river at its outlet from the lake would deprive Egypt of this', and emphasized that the 'occupation of so distant a point as Uganda would be a fair and just claim to render valid our influence in the Nile basin and beyond' (op. cit. Tvedt, 2011: 187).

British banks also had a strong interest in the Egyptian economy. In 1882, Egypt's foreign debt had increased to UK£100 million with an annual debt amounting to UK£5 million, of which a large part went to Britain. The 'Nile water awareness' in London was so great that *The Times* reported regularly on the Nile's water discharges (Tvedt, 2011: 177). The British administration under Lord Cromer understood these links from the very outset and British policy was structured around the Nile. In his imperial strategy, Cromer wrote: 'When, eventually, the waters of the Nile, from the Lakes to the sea, are brought fully under control, it will be possible to boast that Man, in this case the Englishman, has turned the gifts of Nature to the best possible advantage' (op. cit. Tvedt, 2011: 182). When Cromer argued in favour of the construction of the Aswan Dam, he wrote that the dam is of 'utmost importance' because 'the prosperity of Egypt depends wholly on the Nile'. In a letter to Prime Minister Salisbury, he emphasized again: 'There can be no doubt that the most crying want of the country at present is an increase in the water supply' (op. cit. Tvedt, 2011: 183). The first Aswan Dam was completed in 1902, but it was too small to meet the increasing demands of water. The Aswan High Dam was inaugurated in 1971.

Historically, the idea of damming the Nile started as part of the search for the Nile sources. The idea seems to first have been brought forward by Samuel W. Baker in 1868 in *The Nile Tributaries of Abyssinia* where he argued for several dams from Aswan to Khartoum (Baker, 1868). Henry Morton Stanley writes in 1878 with regards to the source of the White Nile at the outlet of Lake Victoria:

> I thought what an immense difference might be effected in the level of the Victoria Falls if Mtsea [the Buganda king] were assisted by engineers. He might by a couple of charges of dynamite, and the destruction of the two islands between which the surplus water rushes into the Victoria Nile, reduce the lake by 12 feet; or he might, by employment of the vast labour at his disposal, dam up the gaps which nature has created, and soon extend the lake thousands of square miles!
>
> (Stanley, 1878: 390)

Winston Churchill too wanted a dam at the source in Uganda: 'It is possible that nowhere else in the world could so enormous a mass of water be held up by so little masonry. Two or three short dams from island to island across the falls would enable, at an inconceivably small cost, the whole level of the Victoria Nyanza' (Churchill, 1909: 120). Owen Falls Dam was inaugurated in 1954.

The 1929 agreement was partly the result of developing the irrigation schemes on the Gezira plain in Sudan, comprising about 5 million acres that would become the world's largest cotton farm, profiting Lancashire. However, if Sudan was going to develop, it needed more water, and quarrels over the Nile with Egypt were unavoidable. The Sennar Dam providing water to the Gezira Scheme was completed in 1925 (Tvedt, 2004: 105–106). The use of Nile water was fundamental to Britain's colonial policy to control Egypt. On 22 November 1924, Lord Allenby announced his famous Nile ultimatum: 'The Sudan Government will increase the area to be irrigated in the Gezira from 300,000 feddans to an unlimited figure as need may require' (Tvedt, 2004: 110). This statement infuriated the Egyptian public and Allenby was asked to revoke the ultimatum. After an exchange of notes, the British government committed itself in guaranteeing Egypt's future water supplies by compensating for the water taken at Sennar, leading to the Nile Waters Agreement (NWA) of 1929.

Controlling the Nile was thus of vital importance for Britain, and the British decision to occupy the Upper Nile can be seen as an example of a far-sighted imperial expansionist policy, driven by a complex mixture of economic and political considerations framed by the geographical and hydrological characteristics of the Nile, according to Tvedt, who further argues that the British had 'become rulers of a truly hydraulic society, where stability and wealth depended upon the water of the Nile' (Tvedt, 2010b: 82).

With the Aswan High Dam inaugurated in 1971 having a storage capacity of two annual floods, water security was controlled within Egypt's borders. Still: 'Although the dam changed the Nile's behaviour in Egypt, it did not liberate the country and the political actors from the power of the structure of the water system itself. The dam made the Egyptians more dependent on the Nile than ever before' (Tvedt and Coopey, 2010: 16). If 'increased and improved water control was destined to top the agenda of any administration in Egypt' by the end of the nineteenth century, the same is true today, but for all riparian states in various degrees (Tvedt, 2004: 22).

The structure of the book: from the sea to the sources

The chapters in this book largely follow the river itself, upstream from the Mediterranean Sea to the lake sources in Uganda and Ethiopia. In Chapter 2, Emil Sandström provides an overview of the scope and magnitude of large-scale land deals taking place in Egypt, Sudan, South Sudan, Ethiopia and Uganda and discusses how the land deals can be understood within the transboundary water management context of the Nile Basin. The chapter discusses how particularly the Egyptian land deals can be understood and how they are intertwined in contemporary negotiations over the allocation of water between riparian states, ongoing hydropower projects and demographic engineering. Sandström suggests that the hydropolitical landscape of the Nile River Basin is about to shift, from a hydropolitical landscape dominated by an Egyptian 'military discourse' towards a 'business discourse', in which new assemblages of actors become shareholders of the Nile waters. Considering the substantial size of some of these land and water deals, actors from countries outside the basin may become important 'hydropolitical players', involved in the future negotiations of the Nile. The chapter concludes that conflicts over land and water resources in Nile Basin region may increase within countries, but not necessarily increase between countries.

While Sandström provides an overview of contemporary land deals in five of the basin countries and discusses Egyptian responses in relation to the changing hydropolitical landscape of the Nile, Kyungmee Kim and Anders Jägerskog (Chapter 3) focus on the investments in Egypt, Sudan and Ethiopia from the nearby Middle East and North Africa (MENA) region (primarily Gulf states). The chapter also provides a critical discussion of the various 'types' of water being used in the basin for agricultural production in the land investments. Is it national or transboundary? Is it blue or green water? The chapter concludes that the investments from the MENA investing countries can be seen as a risk-mitigating strategy in order to achieve a higher level of food and water security. However, such risk-mitigation may be inversed as seen from a Nile Basin perspective where a number of hydropolitical questions can be raised. For example, will political relations between the investing countries and the hosting countries become stronger? Will some of the Gulf countries become indirect hydropolitical actors in the Nile Basin?

In Chapter 4, Harry Verhoeven brings additional complexity by providing a historical overview of how Gulf Arab states have been co-constitutive of the turbulent geo-politics of the Nile Basin for decades and how they have affected both the development of agricultural and water resources as well as struggles over domestic and international hegemony in the Nile. The chapter details with how Ethiopian relations with the Gulf states have been strained for most of the past 60 years almost regardless of governments in Addis Ababa, and how regimes in Egypt and Sudan intermittently have been isolated from the Arabian Peninsula, only to reverse the situation a few years later and to draw on Gulf Arab financial resources and politico-military backing to secure domestic hegemony and expand their regional influence. In recent decades, large hydro-agricultural initiatives in Sudan and Egypt have not just depended on Gulf funding but the very ability of secular, military and Islamist regimes to remain hegemonic in their societies, and the region has been moulded by relations with Riyadh and, more recently, Abu Dhabi and Doha.

In Chapter 5, Martin Keulertz elaborates on a specific case of inward investment in the Nile Basin – namely, that of Qatari plans for investments in Sudan. The chapter analyses the domestic political scene in Qatar and the rationale for them to invest in agricultural production in Sudan as a strategy to achieve water and food security. While much analysis is focused on the countries where the investments are made, Keulertz shows the internal power struggles in Qatar as being co-constitutive of the investments, where different parts of the Qatari shadow state vied for influence over the investments. Disputes within Qatar, however, led to the gradual demise of the investment plans before they actually materialized. Keulertz shows how transboundary water politics is often trumped by food politics, where the interests of states to gain access to capital from large companies and sovereign wealth funds (SWFs) take precedence over transboundary relations.

In Chapter 6, Ana Elisa Cascão and Alan Nicol focus on the changing and specific role Sudan has in contemporary Nile Basin hydropolitics. They situate the discussion on land investments in the current development in the Eastern Nile where the construction of the Grand Ethiopian Renaissance Dam (GERD) in Ethiopia is likely to provide Sudan with more opportunities to use the Nile waters for irrigation which can also improve opportunities to attract land investments. While Sudan traditionally has sided with Egypt in Nile geopolitical matters, it is now clear that Sudan carves out a strategy for itself. Cascão and Nicol show how investments relating to hydraulic infrastructures and expansion of large-scale irrigated agriculture is affecting how Sudan approach negotiations and dialogue on Nile waters.

In Chapter 7, Ramy Lotfy Hanna provides yet another perspective on Sudan by approaching it from a slightly different perspective. Moving beyond the *state-centric* discourse in transboundary water resources politics, the contribution focuses on the role of state enterprises and non-state actors in large-scale land acquisitions (LSLAs) and water grabs in the Nile River Basin. Hanna discusses the hydropolitical relations between Egypt and Sudan and how the governance of the Nile as a transboundary river basin is seeing new actors such as SWFs, state-owned companies or the private sector being more active. However, these new actors are not involved formally in the discussions pertaining to the hydropolitics in the Nile Basin despite being important players.

Where the White Nile meets the Blue Nile in Khartoum, Chapter 8 proceeds to Lake Victoria (the historic source of the White Nile). In this chapter, David Ross Olanya analyses the expansion of dam constructions in Uganda. Although there are huge land acquisitions in Uganda, both foreign and domestic, these are predominately on rain-fed agricultural land and it is therefore difficult to assess to what extent it may, if at all, affect the overall amount of water in the basin. Dam constructions in Uganda are part of nation-building and a social transformation of the economy from a peasant economy to a middle-income country by 2040. While dam constructions inevitably involve displacement of people and internal activism as well as opposition from foreign non-governmental organizations, in the overall Nile Basin geo-politics these hydroelectric dams receives little attention. As a case contrasting other developments along the Nile north of the Sudd, nation-building structured around water and hydropower may fulfil internal national purposes in Uganda while at the same time not harnessing the interests of downstream users since the constructed dams are in rather narrow gorges.

One of the tributaries to the White Nile is the Sobat or Baro-Akobo River coming from Ethiopia. On the Ethiopian side of the border to South Sudan is

Gambella – a massive area of land investments utilized for both irrigated and rain-fed agriculture. In Chapter 9, Wondwosen Michago Seide analyses the conflictual context of the land investments in the Gambella region. As one of the most marginalized regions in Ethiopia, land is leased almost for free but, despite initial low costs of the land itself, it has turned out to be very difficult to operationalize the land investments. Unpredictable water regimes – floods and heavy precipitation – but also malaria, difficult transport of goods and not least violent conflicts between the Nuer and Anywaa, make the area less optimal for large-scale agricultural investments than anticipated despite an abundance of water and land. More than other cases, this chapter shows that 'water' and 'land' grabbing is much more than abundance of water and land, and in this context Ethiopian officials do not see investors as 'grabbers' but 'development heroes' taming nature and benefitting the country and nation.

Moving further up in the Ethiopian highlands Mats Hårsmar, Emil Sandström and Atakilte Beyene (Chapter 10) explore past and present hydropolitical ambitions along the Ethiopian Blue Nile with a focus on what the Ethiopian government has pronounced as a 'growth corridor' area around the Lake Tana area (the source of the Blue Nile). The chapter provides a historical review of Lake Tana's central role in the Nile Basin's hydropolitical history and describes a number of ongoing and planned hydraulic projects in the Lake Tana area, which are discussed in relation to Ethiopia's broader national development efforts and changing hydropolitical relations with its riparian neighbours. The chapter argues that the contemporary developments on the Blue Nile and in the Lake Tana area needs to be understood in relation to both past and present hydropolitical trajectories and concludes that contemporary developments along the Nile, with the signing of the CFA, a radically stronger Ethiopian economy, political turmoil in Egypt in combination with huge potentials for silt reduction in the Lake Tana area have opened up a leeway for Ethiopia to go ahead with her long-held ambitions to develop the Blue Nile waters for irrigation and hydropower.

In Chapter 11, Terje Oestigaard focuses on yet another but little emphasized theme, namely how national identities are not only structured around safe guarding the life-giving waters, but also very much shaped by the very same water. By comparing identities of water in Egypt and Ethiopia, century- and even millennia-long trajectories and traditions have political implications today. From Herodotus's time, Egypt has been perceived as the 'gift of the Nile'. Gish Abay – the source of the Blue Nile in Ethiopia – is perceived as the divine outlet from Paradise, and for a millennium religious beliefs in Egypt and Ethiopia have held that the Christians in Ethiopia may block or divert the Nile from the Muslims in Egypt. National water identities are part of shaping geopolitics, and threats about water wars have been prevalent in the political discourse between Egypt and Ethiopia. Whether this will take place is yet another question, but water as a national ideology is a very powerful tool when used to mobilize groups or nations against others.

Emerging trends and challenges

What emerges from the research in this book are at least two wider trends relating to the contemporary hydropolitical landscape of the Nile Basin, namely increased complexity and the absence of a water hegemon with the ability to control other riparian countries' water use. Several of the chapters in this volume discuss the changing role of Egypt as the principal water hegemon in the Nile Basin. The signing of the CFA, recent political turmoil in Egypt and the entering of new assemblages of land investors, in combination with a stronger Ethiopian economy, the split-up of Sudan and a civil war in South Sudan, etc., have in various ways contributed to a shift in the hydropolitical landscape in the Nile Basin. Although it is common to talk about hydro-hegemony today in academic circles, in the current geo-political situation it seems as if there is no strong hegemon in the Nile Basin. From the several historical reviews in this volume, it could even be argued that neither Egypt nor Britain have had full hegemony. Egypt and Britain may have strived for hydro-hegemony, but they have never really been able to exercise it in full.

Given that there are 11 independent states in the basin and numerous foreign interests involved in water and land acquisitions, the geo-politics are, in practice, fluid and in flux. Due to the extremely complex and changing situation, it is thus difficult to conclude with any specific geo-political consequences or implications on the implicit and explicit water uses in the land acquisitions. The complexity and the lack of any clear hydro-hegemon may enable unprecedented and unforeseen developments in the Nile Basin since there are so many actors and countries, within and beyond the basin, acting individually or in cooperation, having partly mutual and overlapping, but often contradicting, interests.

With the diminishing role of the prevalent dominating state in the basin (Egypt), it seems in practice that most riparian states go ahead with their water developments unilaterally although being politically and rhetorically framed in accordance with the principles of the CFA and the NBI. As in the past, the agricultural question remains crucial in these developments, particularly when considering the expected population growth in the basin and recurring droughts and famines. The very absence of clear data on, for instance, water withdrawals also challenges the concept of hydro-hegemony, since the realities on the ground differ from political discourses and investment plans. As such, the investments in land become even more challenging, related to agriculture and food security as they are.

What emerges from the different chapter contributions are thus a complex web of relations that are primarily political but also increasingly commercial, and need to be further analysed in order to better understand their subsequent hydropolitical ramifications. Since the Nile does not follow national borders, and investors interested in water and land are neither strictly limited by nation states (although the deals take place according to different national laws), an investor may engage in different Nile Basin countries at the same time but on various terms. Still, the Nile connects all and water utilization upstream may affect downstream countries; hence, foreign investors may sometimes intimately be interwoven in diametrical relations, sometimes even in opposition to national geo-political interests in the Nile Basin, particularly with regard to Egypt, Sudan and Ethiopia.

To some extent one could argue that stakes are higher today than in the past, because it is not only about capitalism and profit. Rapid population growth in the whole basin will put more pressure on the basin's water and food security needs when foreign governments and corporations are scrambling to secure control over food supply chains. This 'double' interest of a country (e.g., Saudi Arabia) – both in relation to requesting a decent turnover of its investments as well as for fulfilling its own domestic food security needs – may turn some countries outside the basin into indirect hydropolitical players in future Nile Basin politics. Complicated as it is today with 11 riparian countries, the added layers of investors (countries, companies, SWFs, etc.) dramatically increases the hydropolitical complexity in the basin. The issues that will define the future of the river basin thus largely remain: under what circumstances could the 11 sovereign states – some water poor, others water rich, some upstream other downstream – agree to manage the common waters of the Nile for the common good?

Notes

1 The views expressed by Jägerskog do not necessarily reflect the views of the Swedish International Development Co-operation Agency or the Swedish government.
2 Hitherto, the transboundary water dimension of the current land investments have largely been overlooked in the literature, except for a few original contributions (see, e.g., Jägerskog *et al.*, 2012; Mehta *et al.*, 2012; Warner *et al.*, 2013).
3 The 1929 agreement with Britain – representing its then East African colonies – gave Egypt the right to veto upstream projects that would affect its water share.
4 Julius Nyerere, the first president of Tanzania, considered that international agreements dating from colonial times should be renegotiated when a state becomes independent, as the nation should not be bound by something that the nation was not in a sovereign position to agree to at that time.

References

Adar, K. G. (2011). Kenya's foreign-policy and geopolitical interests: The case of the Nile River Basin. In Adar, K. G. and Check, N. A. (eds). *Cooperative Diplomacy, Regional Stability and National Interests: The Nile River and the Riparian States*: 167–188. Pretoria: Africa Institute of South Africa.

Baker, S. W. (1868). *The Nile Tributaries of Abyssinia, and the Sword Hunters of the Hamran Arabs*. Philadelphia: J. B. Lippincott and Co.

Borras, S. M. and Franco, J. C. (2013). Global land grabbing and political reactions 'from below'. *Third World Quarterly*, 34(9): 1723–1747.

Cascão, A. (2012). *Hydropolitics in the Nile Basin: Water, Power and Cooperation in the Post-Colonial Era*. London: I.B. Tauris.

Cascão, A. and Zeitoun, M. (2010). Power, hegemony and critical hydropolitics. In Earle, A., Jägerskog, A. and Ojendal, J. (eds). *Transboundary Water Management: Principles and Practice*: 27–42. London: Earthscan.

Churchill, W. S. (1909). *My African Journey*. Toronto: William Briggs.

Cotula, L. (2013). *The Great African Land Grab? Agricultural Investments and the Global Food System*. London: Zed Books.

Edelman, M., Oya, C. and Borras, S. M. (2013). Global land grabs: Historical processes, theoretical and methodological implications and current trajectories. *Third World Quarterly*, 34(9): 1517–1531.

Jägerskog, A., Cascão, A., Hårsmar, M. and Kim, K. (2012). *Land Acquisition: How Will it Affect Transboundary Waters?* Report No. 30, Stockholm: SIWI.

Kendie, D. (1999). Egypt and the hydro-politics of the Blue Nile River. *Northeast African Studies*, 6(1–2): 141–169.

Mehta, L., Velderisch, G. J. and Franco, J. (2012). Introduction to the special issue: Water Grabbing? Focus on the (Re)appropriation of Finite Water Resources. *Water Alternatives*, 5(2): 193–207.

Muhammad, J. H. (2010). The dams of Northern Sudan and the policy of demographic engineering. *International Journal of African Renaissance Studies*, 5(1): 148–160.

NBI (2013). *The Nile Basin Initiative*, http://nilebasin.org (accessed December 2015).

Salman, M. A. (2011). The new state of South Sudan and the hydro-politics of the Nile Basin. *Water International*, 36(2): 154–166.

Stanley, H. M. (1878). A geographical sketch of the Nile and Livingstone (Congo) Basins. *Proceedings of the Royal Geographical Society of London*, 22(6) (1877–1878): 382–410.

Tvedt, T. (2004). *The River Nile in the Age of the British: Political Ecology and the Quest for Economic Power*. London: I.B. Tauris.

Tvedt, T. (ed.). (2010a). *The River Nile in the Post-Colonial Age*. London: I.B. Tauris.

Tvedt, T. (2010b). Water: A source of wars or a pathway to peace? An empirical critique of two dominant schools of thought on water and international politics. In Tvedt, T., Chapman, G. and Hagen, R. (eds). *A History of Water, Series 3, Vol. 3: Water, Geopolitics and the New World Order*: 78–108. London: I.B. Tauris.

Tvedt, T. (2011). Hydrology and empire: The Nile, water imperialism and the partition of Africa. *Journal of Imperial and Commonwealth History*, 39(2): 173–194.

Tvedt, T. and Coopey, R. (2010). A 'water systems' perspective on history. In Tvedt, T. and Coopey, R. (eds). *A History of Water, Series 2, Vol. 2: From Early Civilizations to Modern Times*: 1–26. London: I.B. Tauris.

UN (2009). *World Population Prospects: The 2008 Revision*: 6–10. Nairobi: UN.

Warner, J., Sebastian, A. and Pinotti, E. V. (2013). Claiming (back) the land: The geo-politics of Egyptian and South African land and water grabs. In Allan, J., Keulertz, M., Sojamo, S. and Warner, S (eds). *Handbook of Land and Water Grabs in Africa*. London and New York: Routledge.

Waterbury, J. (2002). *The Nile Basin: National Determinants of Collective Action*. New Haven and London: Yale University Press.

Whittington, D., Wu, X. and Sadoff, C. (2005). Water resource management in the Nile Basin: The economic value of cooperation. *Water Policy*, 7: 227–252.

Zeitoun, M. and Mirumachi, N. (2008). Transboundary water interaction I: Reconsidering conflict and cooperation. *International Environmental Agreements*, 8: 297–316.

2 Dealing with water

Emerging land investments and the hydropolitical landscape of the Nile Basin

Emil Sandström

Introduction

Much of the recent literatures on 'land grab' have attributed the surge of interest in land to a number of linked global crises related to finance, food, fuel (peak oil) and climate that emerged in 2007–2008 (Cotula *et al.*, 2009; Friis and Reenberg, 2009; McMichael, 2010; De Schutter, 2011). Countries with relatively cheap prices for land and labour are markedly seen as attractive destinations for foreign investments in agriculture. Increasing evidence, however, indicates that control of water resources is an important cause and effect of the phenomena commonly known as land grabbing. It has been observed and argued that water is both a target and a driver of this phenomenon (Mann and Smaller, 2010; Woodhouse, 2012; Mehta *et al.*, 2012; Franco *et al.*, 2013). Accurate information on the magnitude of land deals and their implications for transboundary water management and water availability is, however, limited. Although land and water are inextricably connected, water has generally been ignored in much of the literature on land grabbing (Mehta *et al.*, 2012) and issues of land grabbing have rarely been linked to transboundary water issues, despite the obvious fact that many of the land deals will have downstream effects.

By framing the current land deals within the hydropolitical landscape of the Nile, this chapter seeks to contribute to the discussions on how land deals can be understood. A focus on land deals and hydropolitics may help to bring out an additional set of issues that relate to transboundary water dilemmas and to the phenomena commonly known as land grabbing. More specifically, the chapter explores the scope and magnitude of large-scale land deals taking place in Egypt, Sudan, South Sudan, Ethiopia and Uganda over the period of 2008–2013 and discusses how the deals can be understood and how they may influence possibilities for conflicts in the transboundary water management context of the Nile Basin. The five riparian countries have been selected on the basis that they are often mentioned in the literatures among the five of the 11 countries in the Nile Basin where the stakes and interests are considered as particularly high (c.f., Waterbury, 2002; Salman, 2011).

The chapter discusses how Egyptian land deals can be understood and how they are intertwined in contemporary negotiations over the allocation of water between riparian states, ongoing hydropower projects and demographic engineering. The chapter suggests that the hydropolitical landscape of the Nile River Basin is about to shift, from a hydropolitical landscape dominated by an Egyptian 'military discourse' towards a 'business discourse', in which new assemblages of actors become shareholders of the Nile waters. Considering the substantial size of some of these land and water deals, actors from countries outside the basin may become important 'hydropolitical players' and be involved in the future negotiations of the Nile.

Unlike many other transboundary rivers systems in the world, the downstream riparian Egypt has often been measured as the water hegemon in the basin due to its military and economic domination (Cascão and Zeitoun, 2010). After the British withdrew from Nile Basin affairs in the 1950s, Egypt basically continued the British tradition to dominate the discourse on the Nile Basin, and has by and large been able to maintain its allocations and water use by withholding to the 1929 and 1959 agreements of which the latter agreement entitles Egypt to the lion's share of the Nile water.[1] Water occupies a central concern in Egypt's foreign policy, national security and military planning.[2] Backed up with a strong position in the region, politically and militarily, Egypt has over the years opposed the establishment of hydropower and irrigation in the basin and has on several occasions threatened to go to war if the precious Nile water is perceived in danger.[3]

The first part of the chapter briefly presents the methods used for the study and some of its conceptual underpinnings. The second part provides an estimate of the land deals carried out in the basin over the period 2008–2013 and examines the implication for water availability and access. Part two also provides a more detailed overview of Egypt's land (water) investment efforts in upstream countries. Section three explores how the land deals and particularly the Egyptian land deals can be understood in relation to ongoing events in the Nile Basin, with contemporary negotiations of the Nile, large hydro-power developments, demographic engineering and the possible emergence of new alliances with actors from countries outside the basin. Section four discusses some of the processes that unfold when land deals are materialized and the implications for customary tenure, rural livelihoods and conflicts. The fifth and final part summarizes and discusses in what ways the land deals may influence conflicts in the transboundary water management context of the Nile River Basin.

Conceptual framing and methodology

The Nile shares in some respects the characteristics of a classical commons dilemma (i.e., it is difficult to exclude others from resource use and resource use affects the possibility for others to use the same resource). However, the Nile River cannot be regarded as a pure commons dilemma, similar to what Hardin (1968) once described on a pasture common. A pasture common or a forest common, for that sake, is not comparable to a transboundary river system, since within a river system one riparian can exploit the river (maximize the yield) without negatively harming the other (c.f., Tvedt, 2010). For example, a downstream riparian country can utilize the river without negatively affecting an upstream riparian and upstream water use may not necessarily have negative downstream ramifications. A hydro-electrical dam upstream may, for instance, under certain conditions even generate several positive outcomes for downstream riparian countries by, for example, reducing the amount of silt flowing downstream and preventing floods. Still, there exist a number of dilemmas attributed to transboundary water resources, not least with regard to the Nile Basin, since upstream countries always have the potential to harm a downstream riparian. In that sense transboundary waters inherently reflect power asymmetries that often are tainted and embedded in historical claims from the past.

This chapter is grounded in a perspective of political ecology that seeks to understand the commons dilemma(s) in the Nile River Basin as manifestations of negotiated social practices located both globally, regionally, nationally and subnationally and within wider contexts of history, contemporary hydropolitics and economy (c.f., Berry, 1993; Sandström, 2008; Cleaver, 2012). Framing the analysis in this way makes it possible to elucidate how the land deals are intertwined in current politics, and in discursive struggles of gaining hegemony of the resource system. In such struggles, political and cultural symbols of power are often brought into play in shaping the outcomes of the resource system (c.f., Lund, 2006). Many of these deals can be seen as acts of territorialization (Ribot and Peluso, 2003) where new assemblages of actors are trying to pursue hegemony in ways that Gramsci might have labelled 'the terrain of global conjuncture' (Gramsci, 1971, cited in Peluso and Lund, 2011: 671).

In this chapter the term land deal is mainly used. It refers to land investments that are incipient and that generally and potentially involve the conversion of subsistence agricultural systems to large-scale commercial use – a process that can be referred to as 'commodification of nature' or as a strategy of 'accumulation by dispossession' – where states collude with capital to embezzle nature and the commons (Harvey, 2003; Peluso and Lund, 2011). The term land deal refers to both intended and/or concluded attempts to acquire land through purchase, lease or concession. It is thus used in a broad, but not imprecise sense, and the chapter differentiates between so-called intended and closed deals. Intended deals are land deals that have been announced but not materialized in any form of written contract. They can be seen as statements of intent from businesses and/or governments. Intended deals have been included since they constitute an important source of information for the understanding of the hydropolitics of the Nile Basin. Closed deals are deals with some form of written contract that generally have been registered in official records.

The land deal data is derived from a combination of data sources. The Land Matrix database has been used as an entry point for getting a crude estimate of the scope of the land deals carried out among the five selected countries in the Nile Basin over the period 2008–2013.[4, 5] The database contains announced and realized land deal transactions that entail a transfer of rights to use, control or own land through sale, lease or concession. The dataset records the name of the investor(s), country of origin, geographical position of the deal, size of the deal and crops to be grown. It mainly focuses on land deals that involve foreign investors, although domestic deals also are documented. The data comes from a variety of sources including official records, academic research, and reports by international organizations, non-governmental organizations and the media (Anseeuw *et al.*, 2013). An advantage is that the database differentiates between announced and realized transactions (intended and closed deals). Announced deals are important as they may reveal different actors' hydropolitical intentions.

The Land Matrix has been criticized for relying too much on media sources and that the dataset contain duplicates.[6] In order to circumvent inaccuracies, the data has been systematically screened and triangulated with business reviews from companies that are actively involved in the land deals in combination with journal articles and reports that have carried out countrywide land deal assessments that rely more on official data in combination with their own field studies (Deng, 2011; Bossio *et al.*, 2012; Lavers, 2012a, 2012b; Olanya, 2014; Chapters 7 and 9 in this volume). When triangulating the data, a number of deals that were reported as closed deals in the dataset has been re-categorized as intended deals and a few deals have been added. Land deals below 500 hectares and deals that are not carried out within basin boundaries have been excluded from the dataset as well as deals that are related to nature conservation purposes. For some of the reported land deals it has been difficult to assess the geographical location. These deals have been left out since they are not considered to significantly impinge upon the overall picture due to their relatively small contribution in terms of size. As a result, the data in the following section does not constitute a comprehensive listing of all large-scale land deals among the five selected countries within the Nile Basin. Nonetheless, it does enable a number of observations in relation to water use and hydropolitics, detailed in the subsections below.

Assessing the volume of land deals in the Nile River Basin

Overall picture of land deals in the Nile Basin

During 2008–2013, the riparian states in the Nile Basin were targeted for a number of large-scale land deals. Although there are uncertainties about the statistics, the figures presented in Table 2.1 provide an indication of the magnitude and scope of the intended and contracted land deals within the basin. The available data shows that the total amount of intended land deals among the five riparian countries comprise an area of more than 8.1 million hectares, and the size of contracted land deals encompasses an area of about 4.3 million hectares (1.65 times the size of Rwanda).

Table 2.1 Intended and concluded large-scale land deals from 2008–2013 among five of the riparian states in the Nile Basin[i]

Target country	*Intended deals*[ii] *(hectares)*	*Closed deals (hectares)*	*Number of closed deals*	*Major crops*	*Major investor country*
Egypt	53,752	119,552	6	Rice, wheat and fodder crops	United Arab Emirates (UAE), Saudi Arabia
Ethiopia	2,304,446	1,237,396	69	Cotton, sugar cane, jatropha and various cereals	Saudi Arabia, India, Ethiopia, USA
South Sudan	2,629,000	1,642,584	11	Various cereals and teak	USA, Egypt
Sudan	2,198,842	1,297,529	19	Wheat, maize, sunflower and sugar cane	Egypt, South Korea, Qatar
Uganda	940,363	38,415	6	Wheat and sugar cane	Egypt
Total	*8,126,403*	*4,335,476*	*111*		

Notes

[i] Data on domestic investors are likely to be underestimated. For example, in Uganda, only one domestic investor is reported in the dataset. One reason for the poor records on domestic land deals is the hardship of getting access to domestic land deal data. The data on domestic land deals in Uganda are, for example, stored at district levels whereas the data on foreign investments can be obtained relatively easily from the national records. Another reason is that many of the domestic land deals are not able to meet the criteria of 500 hectares to be considered as a 'large-scale' land deal. Generally, the reported domestic land deals tend to involve smaller land areas than the foreign investments.

[ii] Intended deals refers to deals that have failed or are in the process of being negotiated.

Sources: Adapted from Deng (2011); Bossio *et al.* (2012); Lavers (2012a, 2012b); Olanya (2014); Land Matrix (2015); Chapters 7 and 9 in this volume.

In total the dataset comprises 111 closed deals over the studied period. Although the land deals are reported as closed deals in the dataset, there are questions concerning the viability of several of the deals. Increasing evidence indicates that many of the announced deals have not yet materialized, and there are several ways in which investment projects may fail. First, the land deals may just be a source for speculation (rent seeking) on increased value of the land. Second, investments can be threatened by land disputes or conflict in project areas. Third, the government may cancel the deal. This is particularly likely with investments that are made between private parties without government involvement or if the deals are made without the proper required involvement of community representation. Fourth, companies may go bankrupt or they simply do not have the necessary capacity to engage in large-scale commercial farming (c.f., Deng, 2011; Abdallah *et al.*, 2014). Despite the fact that many of the reported deals have not yet materialized, they are still interesting to study from a hydropolitical point of view, as they can be seen, deliberately or not, as claims that aim to secure control over land and water resources along the Nile.

The vast majority of the recorded investors are private companies. Five governments are directly involved in agricultural investments in the basin: Egypt, the UAE, the Republic of Korea, Saudi Arabia and Djibouti. The two largest reported land deals in the basin are the 640,000 hectares South Korean deal in Sudan and the 600,000 hectares deal in South Sudan, where the US-based Nile Trading & Development Inc. and the Mukaya Payam Cooperative have obtained a 49-year leasehold. A considerable amount of controversy surrounds the latter project and whether this project will ever be operational is an open question (Deng, 2011; Oakland Institute, 2011). Other leading companies in terms of size are the Egyptian shareholder company Citadel Capital, the Indian Karaturi Global Ltd and the MIDROC business empire owned by the Ethiopian/Saudi Billionaire Sheik Mohammad Al-Amoudi.

From the statistics it is possible to discern that South Sudan stands out as the prime target country in terms of comprised size of the deals, followed by Ethiopia and Sudan. It is also clear that the investors behind the land deals are mainly companies from countries outside the Nile Basin (e.g., Saudi Arabia, India and the USA); but interesting to note is that Egypt is one of the leading investors in terms of intended size of the deals in Sudan, South Sudan and Uganda.

Egyptian land deals in the Nile Basin

The dataset reveals that Egyptian investors during the past five years have expressed an interest to acquire 1.7 million hectares in Sudan, South Sudan, Ethiopia and Uganda. About 473,000 hectares are considered as closed deals (see Table 2.2).

Among the Egyptian land dealers are the aforementioned Citadel Capital and the Government of Egypt. Citadel Capital is considered as one of the leading investment companies in Africa controlling investments of US$9.5 billion across several sectors, including energy (oil), transportation, agrifoods and cement (Citadel Capital, 2012). Through its platform companies Wafra and Sabina, it has been able to secure a 30-year lease for agricultural production in Sudan and a 25-year lease in South Sudan. The investments in Sudan comprise an area of 136,080 hectares of which 37 kilometres is located directly on the river banks of the Nile (Citadel Capital, 2012). In north South Sudan, to the border of Sudan, Citadel Capital has been able to obtain a lease of 105,000 hectares. Both projects engage in large-scale cultivation of cash crops, including sugarcane, grain, sorghum, maize, sunflower, rice and various grain legumes. Besides these two leaseholds the company also has expressed intentions to develop large-scale agricultural schemes in Kenya and Uganda (Citadel Capital, 2012).

Table 2.2 Egyptian land deals between 2008 and 2013 in Ethiopia, Sudan, South Sudan and Uganda

Investor	Target country/ location	Intended deals (hectares)	Closed deals (hectares)
Citadel Capital	Sudan		136,080
Citadel Capital	South Sudan		105,000
Government	Sudan	420,000	210,000
Government	Uganda	800,000	
Government	South Sudan	16,800	
Government	Ethiopia[i]		22,000
Total size		*1,236,800*	*473,080*

Note
[i] Located outside the Nile Basin in the Afar region.

The land deals on behalf of the Government of Egypt are handled by the Department of Foreign Agriculture Relations at the Ministry of Agriculture and Land Reclamation. The governmental land deals over the period involve 210,000 hectares in Sudan and 16,800 hectares in South Sudan. In addition to the above land investment efforts, Egypt has also acquired a lease of 22,000 hectares in Ethiopia. This investment, however, is situated in the Afar region outside the Nile Basin area. Ownership patterns of the governmental land deals are unclear, as are the constituent actors, since the contracts are not disclosed to the public. However, what seems certain is that the land investments are supposed to be carried out as Egyptian public–private partnerships, which are primarily intended to ensure food production for Egypt. Besides land deals in Sudan, South Sudan and Ethiopia, the Government of Egypt has also showed interest to invest in Uganda and to take over large parts of the famous Gezira scheme in Sudan.

In 2010, the Sudanese daily papers filed stories that the Sudanese Ministry of Agriculture had signed an agreement in September with the Egyptian Ministry of Agriculture that 400,000 hectares of the Gezira Scheme should be handed over to Egyptian management (*Sudan Tribune*, 2010a). In accordance with a certain contract, combinations of public and private Egyptian corporations were to provide the necessary inputs and agricultural production was to be exported to Egypt. The news aroused protests from the Sudanese Farmer's Union and the tenants in the Gezira Scheme, as well as commentators in the Sudanese media. Their argument was that no activity should take place before the issue of landownership in the Gezira Scheme was settled by the government. Some even described the deal as a betrayal of the Sudanese tenants, farmers and the Sudanese people since the scheme played a significant role in the life of the Sudanese people (*Sudan Tribune*, 2010b). The contract was to be signed in October the same year. However, the signing of the contract was adjourned for the foreseeable future. Some explain that the postponement might be due to the pressures on the Sudanese side not to allow the deal to be executed at the expense of the Sudanese tenants and farmers, while other news commentators argued that Egypt had shifted its land investments focus towards other Nile Basin countries (*Sudan Tribune*, 2010c). Two years later, a new land deal between the Governments of Sudan and Egypt was announced. This time, the land deal concerned 210,000 hectares. The deal was settled between the two countries in connection to a high-level meeting between the presidents of Sudan and Egypt in April 2013 (*Global Post*, 2013).

The Government of Egypt has also expressed an interest in acquiring land in Uganda. In 2008, Reuters reported that the Egyptian media said that Uganda had allocated about 800,000 hectares of land to Egypt to cultivate wheat and corn. Egypt's Minister of Agriculture would together with a delegation of government officials and businessmen travel to Uganda in order to negotiate final details and prepare to start work on the first 200,000 acres (Reuters, 2008). The news aroused a vehement debate among news commentators in Uganda, and while the Government of Uganda denied the deal, Uganda's parliament called an emergency session to investigate the matter. In February 2010, it was reported again that the Government of Egypt would send a delegation of agricultural experts to Uganda to assess farmland suitable for growing wheat to be imported back to Egypt. An adviser to Egypt's Ministry for Agriculture stated to Reuters that the Egyptian agricultural ministry was looking at three options: 'whether Egyptian investors would own and cultivate farmland, whether land would be owned in partnership with Egypt and Uganda, or whether land would be owned as a concession right' (Reuters, 2010).

The issue is not whether the comprised media narratives presented above are 'true' or not. The compiled media story shows how Egyptian land deals in Sudan and Uganda are circumvented with great ambiguities, controversies and debate. Obviously, the Government of Egypt has a strong and persistent interest to develop large-scale agricultural schemes for domestic consumption among the riparian states of the Nile. The media narratives and earlier described land deals indicate that Egypt's hydropolitical strategy is about to shift from a military discourse and strategy that historically has been aimed towards destabilizing upstream neighbouring countries towards a business-oriented strategy that involves land investments in the upstream countries. This strategic change is further substantiated by Egypt's engagement in a project entitled Joint Egyptian Farms in Africa, which has the explicit purpose of the 'realization of Egyptian presence in the African countries' (Egypt's Ministry of Foreign Affairs, 2013).[7] These farms are of a smaller size as the deals discussed above and they are supposed to 'result in enhancing political, economic and inter-trade between Egypt and African countries'. Another aim of the project is to ensure that 'Egypt obtains raw materials necessary for Egyptian industry and agriculture from African countries' (Egypt's Ministry of Foreign Affairs, 2013). To date, the Government of Egypt has been able to establish so-called joint farms in Tanzania, Kenya, Niger and Zambia and negotiations are under way to create a 200 hectare farm in Uganda (Egypt's Ministry of Foreign Affairs, 2013).

Water arithmetic

Water constitutes the backbone of all agricultural activities and agriculture accounts for 70 per cent of freshwater withdrawals from rivers, lakes and aquifers in the world and up to more than 90 per cent in water-scarce countries (UN, 2009). Land deals, however, seldom include water in the contractual agreements (Cotula, 2011), yet the lands that are being negotiated have a significant water footprint, particularly if irrigated. Thus, published information on subsequent water use is not available, but it is likely to assume that a majority of land deals, particularly in the northern part of the Nile Basin, would require agricultural irrigation if they were materialized. De Schutter (2011) has observed that foreign investors tend to focus on areas with good potential of irrigation or areas that receive high rainfall, and he argues that it is unlikely that commercial investments would accept the risks related to unpredictable African rainfall. This is further substantiated by a study by Woodhouse and Ganho (2011) who show in a review of crops proposed for cultivation in land deals in Africa that these crops would require irrigation, because they either intend to cultivate crops with high year-round water requirements (sugar cane, rice) or crops that need irrigation during the dry season (wheat, vegetables, fruit). A similar pattern emerges in the land deal dataset of the five studied countries in the Nile Basin. An overall assessment of the reported crops for agriculture production reveals that it is primarily water-intensive crops such as wheat, rice, sugar cane and cotton that are intended for cultivation (see Table 2.1).

The Food and Agriculture Organization (FAO) estimates that the maximum irrigation potential in the Nile River is 8 million hectares out of which about 5.5 million were irrigated in 2005 (FAO, 2005; Oestigaard, 2012). If only half of the reported intended land deals took place, and 75 per cent of them would use irrigation, the total irrigation potential of the river basin would be exceeded. In practice, this would mean that there are tight water limits for the developments of future irrigation schemes in the Nile Basin from a hydro-ecological point of view, particularly when considering that the above arithmetic does not include land deals in other Nile Basin countries (Burundi, Kenya, Tanzania, Democratic Republic of Congo and Rwanda) and that data on large-scale domestic land deals are considered to be underestimated. Although the data on land deals in the Nile Basin suggest less future quantity of water flowing in the Nile, the land investments may not necessarily significantly hurt the water supply of downstream countries. Water availability fluctuates widely across times and space within the Nile Basin, making it very difficult to predict downstream effects. Furthermore, it is not just a question of water quantity for downstream countries, but also a matter of access, quality, timing and control (Franco *et al.*, 2013). Depending on where, when and which types of crops are to be grown, water quantities used for agriculture could in theory be regulated and adjusted to inter-annual water availability. With the present land deal data at hand it is, however, not possible to pinpoint to what extent such considerations are or even can be made.

Understanding the land and water deal dynamic in the Nile Basin

In order to understand the present land and water deal dynamic along the banks of the Nile and particularly Egypt's surge for land and water resources among its southern riparian neighbours, we need to understand Egypt's special situation in relation to the Nile. The Nile is literally the lifeline of Egypt and few countries are as dependent on food imports as Egypt. Egypt is the largest wheat importer in the world and a large proportion of its wheat imports come from Russia (FAO, 2013). In 2010, Russia and a number of other countries put temporary export bans on agricultural commodities (mainly cereals) due to fear of not being able to secure their own population with enough food (Vassilieva and Smith, 2010). These measures taken by Russia and other food exporting countries contributed to the spiking of food prices in 2010 and represented a risk for Egypt's food security. Egypt's attempts to invest in agriculture in upstream countries can in this light be seen as water insurance policy that could protect Egypt against future food shortages and also is coupled with a projected population increase. The land deals on behalf of the Government of Egypt, however, can not only be understood as solely a food and water security issue for Egypt. There is more at stake. The Egyptian land deals are to some extent arcane, as the agricultural investments at the same time will contribute to reduce the flow of water into Egypt. In other words, Egypt is 'biting its own tail' by investing in agriculture in upstream countries. The land deal attempts on behalf of Egypt, however, appear more logical if they are viewed in relation to contemporary Nile Basin negotiations; the emergence of new assemblages of land deal actors; and the construction of large-scale hydropower in Sudan, Ethiopia and Uganda.

Negotiating the Nile

In 1999, the Nile Basin Initiative (NBI) was established after decades of various United Nations Development Program and World Bank-sponsored projects that aimed to 'encourage sustainable development' in the Nile Basin. The NBI serves as a forum through which member states 'seek to develop the River Nile in a cooperative manner, share substantial socio-economic benefits and promote regional peace and security' (NBI, 2013). Although representatives (ministries of water affairs) of the then nine member countries of the NBI met on a regular basis over the course of the next decade, negotiations did not progress much when it came to how the waters in the basin should be allocated between the countries. Upstream countries had hoped to settle a new agreement that would give them more rights to Nile water than the 1959 agreement, but it soon became clear that Egypt had no intentions of yielding water to other riparian states.[8] In May 2010, Ethiopia, Rwanda, Uganda and Tanzania took an initiative of their own and signed a new water agreement, the so-called Nile Cooperative Framework Agreement (CFA, also known as the Entebbe Agreement). One week after the signing in May 2010, the responsibility for Egypt's negotiations was shifted from the Ministry of Irrigation and Water Resources to the General Intelligence Services headed by Omar Suleiman who later became the vice president of Egypt under the Mubarak regime (Oestigaard, 2012: 39). This shift of governmental affiliation can be interpreted as a symbolic act on the part of Egypt to show the significance of their discontent of the signing of the CFA. Unsurprisingly the CFA was opposed by Egypt and Sudan, who left the negotiations later the same year.

On 19 May 2010, Kenya signed the CFA and when Burundi signed on 28 February 2011, the necessary two-third majorities was reached in order to make the agreement effective after ratification. Although the CFA does not specify how much water is supposed to be allocated between each country and for what purpose, the agreement has symbolic significance for understanding the power and land deal dynamics in the Nile Basin. The CFA has strengthened the upstream countries' position versus Egypt and contributed to a shift of power in the hydropolitical landscape in favour of the upstream countries, where the role of Egypt as the principal Nile water hegemon is not as obvious anymore. The CFA may not in itself change the hegemonic balance prevalent on the Nile watercourse, but it can be seen as a counter-hegemonic move to change Egypt's long-held dominance (Ibrahim, 2011). Furthermore, the Arab Spring and the last years of domestic political turmoil in Egypt have made it unclear if Egypt is able to enforce its role as the principal water hegemon in the region.

In relation to the signing of the CFA in May 2010, the then minister of irrigation and water resources for Egypt (Mohamed Nasreddin Allam) stated: 'The new treaty is not binding to us. It will only be an obligation for the countries that signed it. At the same time we will make sure that this treaty won't affect Egypt's share of the Nile's water' (*Al-Ahram Weekly*, 2010). One way of securing that the treaty will not harm Egypt's water supply to the Nile is to obtain as much land as possible along the banks of the Nile, and from 2010 onwards it is possible to trace most of the Egyptian land deal attempts in Sudan, South Sudan, Uganda and Ethiopia. The Egyptian land deals can, in this respect, be seen as a hydropolitical response and as an act of territorialization (c.f., Ribot and Peluso, 2003; Peluso and Lund, 2011) that aims to prevent other actors' possibilities to gain access to the Nile. By investing in land (water) in the upstream countries, Egypt, on the one hand, assures the availability of water for domestic food production, while, on the other, the country strengthens its presence and influence over the Nile waters in the upstream countries. Egypt's land investment efforts can in this respect be seen as an expression to assert influence over the Nile water course through commercial means. In other words, Egypt is not 'biting its tail' (Nile) but is rather aiming to 'control its tail' through physical presence.

The aspects of presence and control are further substantiated by the earlier mentioned Egyptian 'joint farm' project where one of the main objectives is the 'realization of Egyptian presence in African countries' (Egypt's Ministry of Foreign Affairs, 2013). By acquiring land (water) upstream in diverse agro-ecological zones and geographical locations, Egypt may gain better control over not just the quantity of water flowing downstream, but also increased control over water quality. These bilateral 'farm projects' may also be seen as an effort to exert future diplomatic pressure on riparian officials in order to secure position in the transforming context of the Nile Basin dialogue (or the lack thereof).

New alliances and shareholding patterns

The hydropolitical landscape of the Nile becomes further complicated when new assemblages of actors, often in different kinds of public–private arrangements from countries outside the basin, acquire large tracts of land. It is possible to assume that some of the land deals will create new forms of cross-border alliances and shareholder patterns with regards to the Nile waters. Large investors, governments and companies from Saudi Arabia, Qatar and other water-scarce countries (particularly in the Middle East), but also investors from, for example, the United States, South Korea and India, may become new important 'hydropolitical players' that could influence future negotiations of the Nile.

The secrecy regarding many of these deals, however, makes it difficult to determine to what extent these new assemblages of actors influence the hydropolitical landscape. One actor who does seem to play a role in shaping new shareholder patterns and cross-border alliances is, for example, Sheikh Mohammad Al-Amoudi, an Ethiopian and Saudi citizen who has close links with the Ethiopian government and the Royal Family of Saudi Arabia. Al-Amoudi owns the MIDROC business empire that has carried out a number of large-scale agricultural deals in Ethiopia in recent years.[9] Al-Amoudi is also owner of the Saudi Star Agriculture Development Plc together with the Government of Saudi Arabia. The Saudi Star has invested in a rice farm project, with the objective to export high-quality rice to Saudi Arabia. The company has leased 10,000 hectares for 60 years, and media reports allege Saudi Star's desire to rent an additional 290,000 hectares from the Ethiopian government (Bekele, 2013).

In view of the extensive size of some of the land deals, new assemblages of actors behind large-scale land deals will become, deliberately or not, new shareholders and hydropolitical players in the Nile Basin. For example, in Sudan the Qatari parastatal company Hassad Food have entered into a farmland development joint venture project with the Government of Sudan (75 per cent Qatari ownership) and placed an investment of 100,000 hectares adjacent to the Nile (Keulertz, 2013). The company has established a canal from the Nile to its farmland project that has the potential to convey 1bcm (billion cubic metres) per year to irrigate the farmland. Translated into percentage, the 1bcm per year is estimated to withdraw about a 1.2 per cent share of Nile waters (Keulertz, 2013). This particular investment attempt has been suggested by Keulertz (2013) as being motivated as a means to increase Qatar's geopolitical sphere of interest in the Nile Basin, where Qatar could 'play its preferred role as benevolent mediator … and enable the Gulf State to influence water and food politics in the Nile Basin' (Keulertz, 2013: 347; see also Chapter 6 in this volume). The water share obtained through large-scale land deals by new assemblages of actors can in this light be interpreted as a purchase of a 'water option' to expand different actors' geopolitical influence in the region.

The reconfiguration of shareholders and the creation of new alliances through large-scale agricultural deals thus add new dimensions to the hydropolitical landscape of the Nile that may provide new actors with increased political leverage in Nile Basin hydropolitics. The land deals could in the future impact upon international relations, which could enable Gulf states to influence water and food politics in the highly politicized Nile Basin.

Considering the magnitude of some of the land deals and the subsequent water use, it is reasonable to suggest that the Government of Egypt is pressured to take part in the scramble for agricultural land in the relatively water-rich upstream countries. The logic and arithmetic of the Nile waters can be seen as a zero-sum game that Egypt is determined to partake in. It is partly through this lens that the contemporary land deal dynamics in the Nile Basin and Egypt's persistent attempts to acquire lands in the upstream countries can be understood.

Hydropower and demographic engineering

The contemporary land deals also occur parallel with demographic engineering and with the construction of hydropower. For example, some of the Egyptian land deals taking place in Sudan are argued to be part of politics that are aimed towards an Arabization of the northern region of Sudan which is actively encouraged by the governments of both Egypt and Sudan through tax incentives (Muhammad, 2010). Although the generation of power is the declared purpose of the dams, the fact is that these dams will also be used for large-scale agricultural irrigation and they cause considerable displacement of people (Muhammad, 2010).

Currently, there are at least eight large-scale hydropower projects in different stages of development in the basin.[10] The largest and most controversial is the Grand Ethiopian Renaissance Dam (GERD) in Ethiopia, located on the Blue Nile, close to the border of Sudan, which is expected to produce 6,000 megawatts (MW) of electricity. The reservoir is scheduled to start filling this year (2016) and to be completed by 2017. Besides contributing power it is estimated that the dam has the potential to facilitate the irrigation of about 500,000 hectares of new agricultural lands (Hope, 2012). The dam has raised concerns in Egypt that it will decrease Egypt's water supplies and since the inception of the construction, the dam has been a source of regional controversy that has caused heated debates in the media, and a number of diplomatic controversies. In January 2012, a tripartite committee and an International Panel of Experts (IPoE) was established between Egypt, Sudan and Ethiopia to promote improved understandings of the impacts the project would have on the three countries.

In June 2013 while Egyptian political leaders discussed the IPoE report with former President Mohammad Morsi, it was suggested the military destroy the dam. Without the knowledge of those at the meeting, the discussion was televised live.[11] Morsi's top adviser apologized for the 'unintended embarrassment' and his cabinet released a statement promoting 'good neighbourliness, mutual respect and the pursuit of joint interests without either party harming the other'. However, a few days later, Mohammed Morsi emphasized that he was not 'calling for war'; 'Egypt's water security cannot be violated at all', adding that 'all options are open' (BBC, 2013a, 2013b).

Parallel to the 'hard talks' and diplomatic controversies about the dam construction in Ethiopia and to the dam constructions in Uganda and Sudan, Egypt has been involved in negotiations with these governments to acquire the aforementioned 22,000 hectares of land in Ethiopia, 800,000 hectares of land in Uganda and 420,000 hectares of land in Sudan. It could be argued that Egypt's interest to acquire large tracts of land in Uganda, Sudan and Ethiopia is related to the developments of large-scale hydropower, in which land and water resources can be seen as commodities to be bargained with that could make Egypt look more 'constructively' on the large-scale hydropower projects that are under way in Ethiopia, Sudan and Uganda.

In a similar vein to what is going on in Sudan, although very different, demographic engineering also occurs parallel with large-scale agricultural investments in Ethiopia. For example, in the Gambella region, where the majority of the Ethiopian land deals have taken place, the land deals in the region are part and puzzle of the Ethiopian 'villagization programme' that aims to relocate people into state-created villages. A number of reports from the advocacy groups GRAIN (2012) and Human Right Watch (2012) have accused the Ethiopian government of forcibly relocating about 70,000 people to make room for large-scale land and water investments in the Gambella region. The Ethiopian federal government has, on the other hand, denied that the villagization process in Gambella is connected to the present land deals and they uphold that it is carried out on a voluntary basis (Human Rights Watch, 2012: Appendix IV). Some 1.5 million people nationwide are destined to be resettled in this way, of which 225,000 are from Gambella (Human Rights Watch, 2012).

The above examples illustrate how large-scale land deals are intertwined and occur in tandem with hydropower constructions and demographic engineering that do not just create tensions between countries but more importantly often spur tensions and conflicts within countries.

Contested customary tenure contexts and proliferation of intra-state conflicts

Land and water tenure play an important role of socio-economic differentiation and they are key institutions for mediating social relations and identity in relation to resources (c.f., Berry, 1993; Lund, 2006; Peters, 2013). Issues of large-scale land deals and consequences for rural livelihoods and customary tenure have been widely discussed, especially concerning people's access to land and food security (Cotula *et al.*, 2009; von Braun and Meinzen-Dick, 2009; Matondi *et al.*, 2011; Wily, 2011; Peters, 2013). Generally in the Nile Basin, land tenure systems are customary and rural people rarely hold formal legal rights to the resources they use and depend on. By and large, land and water resources are considered as state property, providing the government with the right to sell and lease the land that in reality is used by farmers and pastoralists. Among the Nile Basin countries, land investments by foreign investors are generally also welcomed by host governments and seen as important contributions to the modernization of these states. Besides policies aimed at resettling farmers within areas served by, for example, large-scale irrigation (LSI) and hydropower, the land investments are spurred by policies that aim to provide a conducive business environment for foreign investors by, for instance, tax reductions (Friis and Reenberg, 2009).

These domestic polices related to land investments in agriculture serve as a vehicle for the enhancement of a privileged elite and at the centre is often a set of contradictory institutions, rules and practices related to how deals are carried out. By and large there are limited legal procedures in place to protect local rights that take account of local interest, livelihoods and welfare (or governments simply oversee these rights) (Peters, 2013). Applications for land are often managed through *ad hoc* procedures at various levels of government, contributing to a lack of transparency and accountability with regard to many of the deals (Beyene and Sandström, 2016).

These policies and ambiguous procedures related to tenure have contributed to the accumulation of land and water resources at the expense of others and prompted a process of polarization between haves and have nots. Although the domestic polices and ambiguous land tenure procedures have not initiated the present surge and accumulation for land and water resources, they have provided the context and the void in which resourceful actors can benefit on the expense of rural livelihoods. As Philip Woodhouse concludes, research now 'indicates that customary tenure acts neither as an obstacle to investment … nor as an inalienable safety net for the poor' (Woodhouse, 2003: 1717, in Peters, 2013). Subsequently, it is not just the common waters of the Nile that are under pressure and threat, but the very idea of 'commoning' as a customary practice that forms the basis for rural people's livelihoods, authority and identity.

Unsurprisingly, there are tensions and conflicts over valued resources as land deals are materialized and disclosed to the public. In 2012, serious protest emerged in the Gambella region (see Chapter 9 in this volume) in one plantation owned by Saudi Star (affiliated company to the MIDROC company) which is irrigated with water that is diverted from the Alwero River (tributary to the White Nile). In April 2012, tension over the project intensified, when an armed group attacked the Saudi Star operation leaving five people dead.

The growing insecurity and violence in Gambella and the loss of human lives can be seen as an indication that intra-state conflicts over land and water resources may well increase where large-scale agricultural investments take place. Similar circumstances involving unprecedented evictions and conflicts in relation to land deals have been reported elsewhere in the Nile Basin. In Sudan, vehement protests have emerged in connection to Egyptian and Qatari land deals in eastern Sudan (Verhoeven, 2013) and in South Sudan prior to the civil war that broke out in December 2013, communities have been caught up in conflicts with a Ugandan business conglomerate called Madhvani Group where they anticipated establishing a 10,000 hectare sugar cane plantation on the river front of the Nile about 70 kilometres north of the capital Juba (Oakland Institute, 2011). In many ways the above examples resemble what emerge in the wake of many land deal cases across the Nile Basin and elsewhere in many parts of the world. There are growing ambiguities, competition, controversies and even armed conflicts among people, state authorities and companies over valued resources (c.f., Wily, 2011; Peters, 2013; Verhoeven, 2013: 52–53). The hydropolitical landscape of the Nile Basin is thus not just a matter of a regional contest between riparian states and outside actors, but as much a locally contested arena, where different claims to resources and resource use are subjected to contests rooted in relations of power at sub-national, national and international levels.

Given that a large number of land transactions have not yet resulted in concrete agricultural investment and production, it is reasonable to expect that land- and water-related conflicts in the Nile Basin will increase as customary production systems are converted to large-scale farms for commercial use. In South Sudan, Sudan, Ethiopia, Kenya and Tanzania, with large populations of pastoralists, there is an even greater risk of conflicts since pastoralist communities require access to large areas of land to graze their livestock.

Conclusions

This chapter shows that beneath the surface of the inflated rhetoric around hydropower projects and how the water should be shared among riparian states, there is an ongoing parallel process, where the riparian states in the Nile Basin have been targeted for a number of large-scale land deals. Estimates reveal that the total amount of intended land deals from 2008–2013 among the five studied riparian countries comprise an area of more than 8 million hectares. The statistics also show that the Egyptian government during the same period has been involved in land deals in upstream countries covering an area of about 1.7 million hectares. Although there are uncertainties about the statistics, the figures presented in this chapter provide an indication of the forces at play as regards the magnitude of the land deals and subsequent water use in the basin.

Although water resources are the backbone of all kinds of agricultural activities, water resources are generally ignored in investment negotiations. Given the scope of the ongoing land deals in the Nile Basin and given that the land deals generally do not include water in the contracts, the waters of the Nile can currently be considered as a 'free for all resource', in which not only actors from the Nile River countries, but also new assemblages of actors that are far beyond the Nile geography scramble to partake. This adds an additional layer of complexity to the water resource dilemma of the Nile. In the worst scenario it could be that the river 'runs short' of water with devastating downstream effects and beyond. Considering the significant size of some of these land deals, actors and governments from countries outside the basin may well become important shareholders and hydropolitical players that could be involved in future negotiations of the Nile. New assemblages of hydropolitical players may redefine relations and influence the power dynamics among the riparian states in the Nile and potentially strengthen weaker states through political support from outside actors. How these new assemblages of actors may impact upon the future negotiations of the Nile is, however, an open question.

The land investment attempts in the Nile River Basin thus relate strongly to the larger question and debate that concerns how the water in the basin should be shared among the riparian states – a debate that has been ongoing for decades and that has been heavily influenced by the military discourse of Egypt. This chapter suggests that the hydropolitical landscape of the Nile River Basin is shifting in light of the large-scale land deals taking place in the region, from a hydropolitical landscape that has been dominated by an Egyptian military discourse towards a business discourse, in which the Government of Egypt along with new constellations of actors compete with each other to purchase Nile water shares. This does not mean that Egypt has completely abandoned its earlier discourse; it has instead moved towards a new kind of practice, where Egyptian attempts to establish large-scale agriculture schemes and so-called 'joint farms' can be seen as an expression to gain influence over the Nile water course. The business discourse through which land and water resources are appropriated is linked to the broader global discourse of the commodification of nature that is made possible by 'the involvement of new capitalist players and actors in water resource management and the rise of new political and economic power relations trough diverse trajectories of neo-liberalism' (Mehta *et al.*, 2012: 198).

The chapter also suggests that the Egyptian land deals are intertwined with contemporary hydropolitics that concern large hydropower constructions, negotiations of a new water agreement and the entering of new assemblages of hydropolitical players in the region. The Egyptian land deals among its riparian neighbours can in these contexts be seen as a hydropolitical response and as a form of territorialization that aims to circumscribe other actors to gain access to the waters of the Nile. Large-scale land deals are thus not only about food, fuel and finance, but also geopolitical struggles over water resources. In this light, large-scale land deals can be seen as a method for gaining geopolitical influence over the Nile water course.

Egypt's attempts to acquire land in the upstream countries can, however, also be seen as an opening that could pave the way for a new Nile agreement, in which all the riparian states could exploit the potential of the river for mutual benefits and avoid a situation leading to a tragedy of the commons. By allowing Egypt to acquire land in the relatively water-rich upstream countries, Egypt might be more willing to alter its resistance to change the old water agreement from 1959 that provides Egypt with the lion's share of the water. From a water-saving perspective, investments in agriculture upstream of the Nile Basin are generally also more favourable than agricultural production in the northern part of the basin, due to less losses to evaporation and the fact that most of the waters from the upper Nile are 'lost' in the Sudd swamps before reaching Sudan and Egypt. Furthermore, it is reasonable to suggest that the more land Egypt has access to in upstream countries for domestic agricultural production, the more implausible it is that Egypt takes up arms against its river neighbours to secure access to the Nile waters in order to feed its own population.

Evidence, however, shows that the present land deals that are taking place in the Nile Basin are likely to spur intra-state struggles over resources and contribute to the accumulation of resources of a landed elite at the expense of rural livelihoods, particularly in situations where people are evicted from their lands and where there are profound uncertainties in land and water tenure. Land and water resources are not just crucial for people's material livelihoods, but also bestowed and embedded in multiple meanings, which are exposed to contests rooted in relations of power both at sub-national and national levels. The sub-national and multiple dimensions of water are imperative factors that so far have often been neglected in the land deals taking place in the basin. Unless more attention is paid to how the recent land and water deals affect people's everyday practices and livelihoods, the Nile Basin will continue to be a volatile region. A preliminary proposition is thus that conflicts and tensions over land and water resources in the Nile Basin region may well increase within countries, but not necessarily increase between countries.

Notes

1 The 1959 agreement stipulates that of the average annual water flow of 84bcm (billion cubic metres) as measured at the Aswan High Dam, Egypt was given the right to use 55.5bcm and Sudan 18.5bcm, leaving 10bcm to evaporation and seapage.

2 The security of the Nile has been central in Egyptian military planning. Egypt's military expenditure in 2005 was more than twice the equivalent expenditure of all the nine other riparian states together (Adar, 2011: 183).

> As deterrence, the Egyptian High Command has established contingency plans for armed intervention, in each country in the Nile Basin, in case of a direct threat to the flow of the Nile. Egyptian military plans, known as *Waraa-elhidoud*, were traditionally associated with Nile water. Some of the plans date back to the early nineteenth century, to the days when Mohammed Ali was rebuilding the Egyptian army. All have been updated several times since then.
>
> (Yacob, 2004: 2)

3 As the late President Sadat stated: 'Any action that would endanger the waters of the Blue Nile will be faced with a firm reaction on the part of Egypt, even if that action should lead to war' (cited in Kendie, 1999). Similar statements have been delivered by the former

presidents Hosni Mubarak and Mohammad Morsi in relation to the developments of hydropower projects, particularly in Sudan and Ethiopia. Political leaders from Ethiopia and other riparian states have also been guilty of making provocative statements.

4 The beta version from 2012 is available at http://landportal.info/landmatrix (accessed December 2013).

5 The Land Matrix is an open public database that aims to promote greater transparency in decision-making over land and investment at a global level by promoting public participation in building a constantly evolving database on large-scale land deals. The Land Matrix is developed and sponsored by the International Land Coalition (ILC), Centre de Coopération Internationale en Recherche Agronomique pour le Développement (CIRAD), Centre for Development and Environment, German Institute for Global and Area Studies (GIGA) and Deutsche Gesellschaft für Internationale Zusammenarbeit (GIZ).

6 See further Anseeuw *et al.* (2013) for a critical discussion of the experiences of using the Land Matrix.

7 See www.far-malr.gov.eg/en/farms.php (accessed November 2013). Link no longer available. In the beginning of 2014 the project seems to have shifted its institutional affiliation from Egypt's Ministry of Foreign Affairs to COMESA (Common Market for Eastern and Southern Africa). Now the project is entitled the Egyptian Farms in Africa. New link www.efa.org.eg/new/en (accessed January 2016).

8 The CFA negotiations primarily revolved around two critical issues. One concerns Egypt's persistence that the country has a historic right to the Nile, acknowledging earlier colonial agreements from 1959. Under Article 14b of the CFA it is stated that the Nile Basin agrees not to 'significantly affect the water security of any other Nile Basin State'. A statement that Egypt and Sudan would like to rephrase to 'not to adversely affect the water security and current uses and rights of any other Nile Basin'. The other disagreement concerns decision-making. Egypt and Sudan want decisions to be taken by rule of consensus, whereas the other riparian states want decisions to be taken by majority vote (Oestigaard, 2012).

9 Besides investments in agriculture, the MIDROC group is also active in mining, construction, tourism, manufacturing, food processing, oil and gas distribution, etc. The MIDROC Investment Group comprises more than 40 companies in Ethiopia. See www.midroc-ethiopia.com/index.html (accessed November 2013).

10 Besides the GERD, Uganda plans to construct three hydropower plants (Karuma, Isimba, Ayago) and Sudan is about to complete the reconstruction of the Roserie Dam and has plans to construct three new dams (Kajbar, Shereik, Dal). See www.internationalrivers.org/campaigns/kajbar-dam-sudan (accessed January 2014).

11 See www.tesfanews.net/morsi-warns-ethiopia-politicians-propose-hostile-acts (accessed January 2016) for the television broadcast.

References

Abdallah, J., Engström, L., Havnevik, K. and Salomonsson, L. (2014). Large-scale land acquisitions in Tanzania: a critical analysis of practices and dynamics. In Kaag, M. and Zoomers, A. (eds) *The Global Land Grab: Beyond the Hype*. London: Zed Books.

Adar, K. G. (2011). Kenya's foreign-policy and geoploitical interests: The case of the Nile River Basin. In Adar, K. G. and Check, N. A. (eds) *Cooperative Diplomacy, Regional Stability and National Interests: The Nile River and the Riparian States*. Pretoria: Africa Institute of South Africa.

Al-Ahram Weekly. (2010). Cairo and Khartoum challenged: Tension over the sharing of River Nile waters is rising. 1–7 July (1005).

Anseeuw, W., Lay, J., Messerli, P., Giger, M. and Taylor, M. (2013). Creating a public tool to assess and promote transparency in global land deals: the experience of the Land Matrix. *Journal of Peasant Studies*, 40(3): 521–530.

BBC. (2013a). Egyptian politicians caught in on-air Ethiopia dam gaffe. 4 June.

BBC. (2013b). Egyptian warning over Ethiopia Nile dam. 10 June.

Bekele, K. (2013). Ethiopia: Saudi Star rice project feels the pinch. *The Reporter*, 23 November 2013.

Berry, S. (1993). *No Condition is Permanent: The Social Dynamics of Agrarian Change in Sub-Saharan Africa*. Madison, WI: University of Wisconsin Press.

Beyene, A. and Sandström, E. (2016). Emerging water frontiers: large-scale land acquisitions and implications for food security in Africa. In Tvedt, T. and Oestigard, T. (eds) *Water and Food: Africa in a Global Context*. London/New York: I.B. Tauris.

Bossio, D., Erkossa, T., Dile, Y., McCartney, M., Killiches, F. and Hoff, H. (2012). Water implications of foreign direct investment in Ethiopia's agricultural sector. *Water Alternatives*, 5(2): 223–242.

Cascão, A. and Zeitoun, M. (2010). Power, hegemony and critical hydropolitics. In Earle, A., Jägerskog, A. and Öjendal, J. (eds) *Transboundary Water Management: Principles and Practice*. London/Washington: Earthscan.

Citadel Capital. (2012). *Annual Report*, http://citadelcapital.com/current-investments/sabina-agriculture (accessed November 2013).

Cleaver, F. (2012). *Development through Bricolage: Rethinking Institutions for Natural Resource Management*. London: Routledge.

Cotula, L. (2011). *Land Deals in Africa: What is in the Contracts?* London: International Institute for Environment and Development (IIED).

Cotula, L., Vermeulen, S., Leonard, R. and Keeley, J. (2009). *Land Grab or Development Opportunity? Agricultural Investment and International Land Deals in Africa*. Rome: IIED, FAO, IFAD.

De Schutter, O. (2011). How not to think of land-grabbing: three critiques of large-scale investments in farmland. *Journal of Peasant Studies*, 38(2): 249–279.

Deng, D. K. (2011). *The New Frontier: A Baseline Survey of Large-Scale Land-Based Investment in Southern Sudan*. Report 1/11, Oslo, Norway: Norwegian People's Aid, www.npaid.org/Media/20_Files/Development/New-Frontier (accessed December 2013).

Egypt's Ministry of Foreign Affairs. (2013). *Egypt's Joint Farm Project*. www.far-malr.gov.eg/en/farms.php (accessed November 2013). Link no longer available. New link, www.efa.org.eg/new/en (accessed January 2016).

FAO. (2005). *Irrigation in Africa in Figures*. AQUASTAT Survey 2005. Rome: FAO.

FAO. (2013). *FAO Statistical Yearbook 2013: World Food and Agriculture*. Rome: FAO.

Franco, J., Mehta, L. and Velderwisch, G. J. (2013). The global politics of water grabbing. *Third World Quarterly*, 34(9): 1651–1675.

Friis, C. and Reenberg, A. (2010). *Land Grab in Africa: Emerging Land System Drivers in a Teleconnected World*. GLP Report No 1, Global Land Project, Department of Geography and Geology, University of Copenhagen, Denmark.

Global Post. (2013). Egypt aims to become self-sufficient in sunflower and oil by using farmland provided by neighboring country. 9 April.

GRAIN. (2012). *Squeezing Africa Dry: Behind Every Land Grab is a Water Grab*. Barcelona: GRAIN.

Gramsci, A. (1971). *Prison Notebooks*. In J. Buttigieg and A. Callari (eds). New York: Columbia University Press.

Hardin, G. (1968). The tragedy of the commons. *Science*, 162(3859): 1243–1248.

Harvey, D. (2003). *The New Imperialism*. Oxford: Oxford University Press.

Hope, B. (2012). Ethiopia's giant dam muddies the waters downstream in Egypt. *The National*, 23 April.

Human Rights Watch. (2012). *'Waiting Here for Death': Forced Displacement and 'Villagization' in Ethiopia's Gambella Region*, www.hrw.org/sites/default/files/reports/ethiopia0112webwcover_0.pdf (accessed October 2013).

Ibrahim, A. M. (2011). The Nile Basin Cooperative Framework Agreement: the beginning of the end of Egyptian hydro-political hegemony. *Missouri Environmental Law and Policy Review*, 18(2): 283–313.

Kendie, D. (1999). Egypt and the hydro-politics of the Blue Nile River. *Northeast African Studies*, 6(1–2): 141–169.

Keulertz, M. (2013). *Drivers and Impacts of Farmland Investment in Sudan: Water and the Range of Choice in Jordan and Quatar*. PhD thesis. Department of Geography, King's College, London.

Land Matrix. (2015). http://landportal.info/landmatrix (accessed December 2015).

Lavers, T. (2012a). 'Land grab' as development strategy? The political economy of agricultural investment in Ethiopia. *Journal of Peasant Studies*, 39(1): 105–132.

Lavers, T. (2012b). Patterns of agrarian transformation in Ethiopia: state-mediated commercialisation and the 'land grab'. *Journal of Peasant Studies*, 39(3–4): 795–822.

Lund, C. (2006). Twilight institutions: public authority and local politics in Africa. *Development and Change*, 37(4): 685–705.

Mann, H. and Smaller, C. (2010). *Foreign Land Purchases for Agriculture: What Impact on Sustainable Development?* Sustainable development innovation briefs. United Nations, Department of Economic and Social Affairs (DESA), Issue 8. New York (January).

Matondi, P., Havnevik, K. and Beyene, A (eds). (2011). *Biofuels, Land Grabbing and Food Security in Africa*. London: Zed Books.

McMichael, P. (2010). Agrofuels in the food regime. *Journal of Peasant Studies*, 37(4): 609–629.

Mehta, L., Velderisch, G. J. and Franco, J. (2012). Introduction to the special issue: water grabbing? Focus on the (re)appropriation of finite water resources. *Water Alternatives*, 5(2): 193–207.

Muhammad, J. H. (2010). The dams of Northern Sudan and the policy of demographic engineering. *International Journal of African Renaissance Studies*, 5(1): 148–160.

NBI. (2013). *The Nile Basin Initiative*, http://nilebasin.org (accessed December 2013).

Oakland Institute. (2011). *Understanding Land Investment Deals in Africa: Country Report: South Sudan*, www.oaklandinstitute.org/understanding-land-investment-deals-africa-south-sudan (accessed October 2013).

Oestigaard, T. (2012). *Water Scarcity and Food Security along the Nile: Politics, Population Increase and Climate Change*. Current African Issues, No. 49. Uppsala: Nordic Africa Institute.

Olanya, D. R. (2014). Asian capitalism, primitive accumulation, and the new enclosures in Uganda. *African Identities*, 12(1): 124–138.

Peluso, N. L. and Lund, C. (2011). New frontiers of land control: introduction. *Journal of Peasant Studies*, 38(4): 667–681.

Peters, P. (2013). Conflicts over land and threats to customary tenure. *African Affairs*, 112/449: 543–562. Oxford University Press on behalf of Royal African Society.

Reuters. (2008). *Egypt to Buy Kazakh Wheat, Grow Grain in Uganda*. 30 August. www.farmlandgrab.org/post/print/2483 (accessed February 2016).

Reuters. (2010). *Egypt and Uganda Mull Wheat Farmland Options*. 15 February. http://af.reuters.com/article/investingNews/idAFJOE61E0LN20100215 (accessed January 2014).

Ribot, J. and Peluso, N. L. (2003). A theory of access. *Rural Sociology*, 68(2): 153–181.

Salman, M. A. (2011). The new state of South Sudan and the hydro-politics of the Nile Basin. *Water International*, 36(2): 154–166.

Sandström, E. (2008). *Reinventing the Commons: Exploring the Emergence of Local Natural Resource Management Arrangements*. Doctoral thesis No. 2008: 48, Faculty of Natural Resources and Agricultural Sciences, SLU. Uppsala: SLU service/repro.

Sudan Tribune. (2010a). Egypt signs deal to allow farmland access in Sudan. 6 September.

Sudan Tribune. (2010b). Hundreds of farmers stage demonstration in El-Gezira state. 10 May.

Sudan Tribune. (2010c). Egypt's takeover of Sudan's Gezira scheme. 17 December.

Tvedt, T. (2010). *The River Nile in the Post-Colonial Age*. London: I.B. Tauris.

UN. (2009). *Water in a Changing World: World Water Development Report*. WWDR3, www.unesco.org/new/en/natural-sciences/environment/water/wwap/wwdr/wwdr3-2009 (accessed December 2013).

Vassilieva, Y. and Smith, M. E. (2010). *Ban on Grain Exports from Russia Comes to Force on August 15*. GAIN Report No. RS1039. Moscow: USDA Foreign Agricultural Service.

Verhoeven, H. (2013). Sudan and its agricultural revival: a regional breadbasket at last or another mirage in the dessert. In Allan, T., Keulertz, M., Sojamo, S. and Warner, J. (eds) *Handbook of Land and Water Grabs in Africa: Foreign Direct Investments and Food and Water Security*. London/New York: Routledge.

von Braun, J. and Meinzen-Dick, R. (2009). *'Land Grabbing' by Foreign Investors in Developing Countries: Risks and Opportunities*. IFPRI Policy Brief 13 No. 13. Washington, DC: International Food Policy Research Institute.

Waterbury, J. (2002). *The Nile Basin: National Determinants of Collective Action*. New Haven/London: Yale University Press.

Wily, L. A. (2011). *The Tragedy of Public Lands: The Fate of the Commons under Global Commercial Pressure*. Rome: International Land Coalition.

Woodhouse, P. (2003). African enclosures: a default mode of development. *World Development*, 31(10): 1717. Cited in Peters (2013).

Woodhouse, P. (2012). New investment, old challenges: land deals and the water constrain in African agriculture. *Journal of Peasant Studies*, 39(3–4): 777–794.

Woodhouse, P. and Ganho, A.-S. (2011). Is water the hidden agenda of agricultural land acquisition in sub-Saharan Africa? International Conference on Global Land Grabbing, Institute of Development Studies. University of Sussex, 6–8 April.

Yacob, Y. (2004). From UNDUGU to the Nile Basin Initiative: An enduring exercise in futility. *Addis Tribune*, 30 January.

3 Overseas investment in land and challenges in the Nile Basin

Evident links from the Middle East and North Africa investment

Kyungmee Kim and Anders Jägerskog

Introduction

The dynamics of land investments in the Nile Basin have been characterized by a search for food security, investment and speculation, as well as regional political influence (see Chapter 2 in this volume). This chapter takes the global discourse of land investments ('grabbing') as a point of departure and focus-specific investments by state actors and non-state actors from the Middle East and North Africa (MENA) in the Nile Basin. While much of the drivers of the investments arguably represent an effort to both increase water and food security for the investing countries, it certainly goes beyond such considerations (Swain and Jägerskog, 2016). Other considerations such as regional political economy relating to political and economic influence are also key considerations. The chapter will aim to outline key considerations of MENA investments in the Nile Basin and its political concerns. Of particular interest will be to try to understand the strategies and choices of the investors (private, semi-private or state) to become 'virtual water hegemons' (Sojamo *et al.*, 2012). Sojamo and others created the concept of 'virtual water hegemony' by combining the virtual water concept and the hydro-hegemony framework (Allan, 1997; Zeitoun and Warner, 2006). The virtual water hegemony framework posits the uneven distribution of power in controlling the flow of virtual water globally.

The chapter takes the global dynamics of land investments in relation to water as a backdrop for an improved understanding of the MENA investments in the Nile Basin and will also initiate a critical discussion of the various 'types' of water that are being used in the basin for production in the land investments. Arguably, distinguishing between various sources of water (blue, green, grey water; water *within* the basin or outside, etc.) is important as that will affect the outcomes of any investments (whether by MENA countries or by any other investors). In this chapter we argue that the increased land investments by MENA countries can be seen as a risk-mitigating strategy that enables them to have more control over their own food security.

Global considerations relating to food security and land acquisitions

Land is an essential and fundamental resource that cannot be replaced. The fundamentality and integrity of land to human life is closely linked to its capacity for agricultural production. Food production is an essential livelihood for farmers, and the importance of food security cannot be overlooked when outlining key considerations of the land investments by MENA countries in the Nile Basin. With a growing population predicted at 8.1 billion by 2025 and 9.6 billion by 2050 from 7.3 billion in 2015, food security considerations will be of increasing importance. The Green Revolution during the 1970s focused on improving crop yields, which significantly increased the productivity in agriculture. However, the adverse environmental impacts from intensive farming and socio-economic limitations dashed the hope of ending the dystopian, Malthusian future through the Green Revolution (Brown and Black, 1970; Conway and Wilson, 2012). Food security is a multifaceted issue that needs to be tackled from both supply and demand sides, including focusing on the supply chain from the field to the consumer, where major losses of food (and therefore also water and energy) is occurring (Lundqvist *et al.*, 2015). Large-scale land acquisitions mainly play a role in the supply side of food security, including water and land availability for food production. Availability of land, water and energy for functioning irrigation systems are key aspects for improving food security. Considerations relating to food security 'at home' by foreign countries investing in the Nile Basin are important when trying to understand the dynamics of how land investments develop. One example relates to Saudi Arabia which decided five years ago that its massive large-scale experiment of growing wheat in the desert irrigated by fossil water (which at one point made them the world's fifth largest wheat exporter) was to be closed down and that they achieved food security through importing wheat instead of producing it at home (Swain and Jägerskog, 2016).

Another key factor to consider is the impact of global climate change to food security. The Intergovernmental Panel on Climate Change highlights that all aspects of food security are potentially affected by climate change, including food access, utilization and price stability; and the effects of climate change on crop and terrestrial food production are already evident in many regions (Porter *et al.*, 2014). Global climate change is predicted to increase the frequency of droughts in dry regions by the end of the twenty-first century (Jiménez Cisneros *et al.*, 2014) and increase water stress in already water-scarce regions. Even the perception of climate change will lead states to become more concerned about water and food security, which already to some extent is evident in the MENA region (Earle *et al.*, 2015; Jägerskog and Kim, 2016). Predictions clearly point to a decreased ability to grow food in the MENA region in the decades ahead (Swain and Jägerskog, 2016). Coupled with global climate change, the global water crisis adds another dimension to the looming food insecurity by causing more competition for water, which provides a rationale to search for food (and water) security elsewhere, such as in the Nile Basin.

The current status of water stress and food insecurity is already alarming among the Nile Basin countries, and the future appears to be more challenging due to the fast-growing population in the basin. Excluding Egypt, all countries in the Nile Basin have been experiencing serious hunger and food shortages in the past decade (IFPRI, 2015). The UN (2015) estimates that the comprised population among the Nile Basin countries will be one billion by 2050 compared to the 487 million in 2015.[1] The predicaments of feeding its growing population will increasingly put pressure on governments in the Nile Basin. The growing commercial and strategic interests in farmland in the Nile Basin are thus likely to worsen the conflicts between land users.

A surge of land deals after the global food price crises

Since 2008, shocks in food prices and threats to global food security have emerged as a major issue in the international policy agenda for development. In 2008, the United Nations established a High-Level Task Force on the Global Food Security Crisis. The aim of the task force was to coordinate UN specialized agencies and other major international organizations and to promote comprehensive and unified responses on global food security.[2] High-level political forums such as the G8 and G20 have been addressing the challenges on achieving global food security and promoting various initiatives on tackling global hunger and ensuring accountability in land deals.

The Food Price Index (see Figure 3.1) shows rapid increases in food commodity prices in the global market during 2007–2008 and 2011–2012. The food price spikes coincide with the growing numbers of large-scale land deals, which have been observed mainly in Africa, Asia and Latin America (Deininger and Byerlee, 2011). These acquisitions involve outright land purchases or, more commonly, long-term leases mainly on government-owned land. Combined with water scarcity in the Middle East, North Africa and parts of Asia, this has led to increased international and domestic investments in agricultural land, primarily in Africa and Latin America (Deininger and Byerlee, 2011), which has raised a series of concerns relating to issues such as land and water rights, social and economic equity, and environmental sustainability. The governments of developing countries where land acquisitions are taking place intend to modernize their agricultural sector and to provide employment in rural areas (Woodhouse, 2012b). However, concerns about social, economic and environmental sustainability raise questions on the intention of investment-receiving governments.

The scale of land acquisitions is alarmingly significant. The Land Matrix, a database of global land deals, presents the information on over 1,000 concluded deals covering 37.8 million hectares.[3] Initial figures on the global scale of land acquisitions have been estimated over 83.2 million hectares (Anseeuw *et al.*,

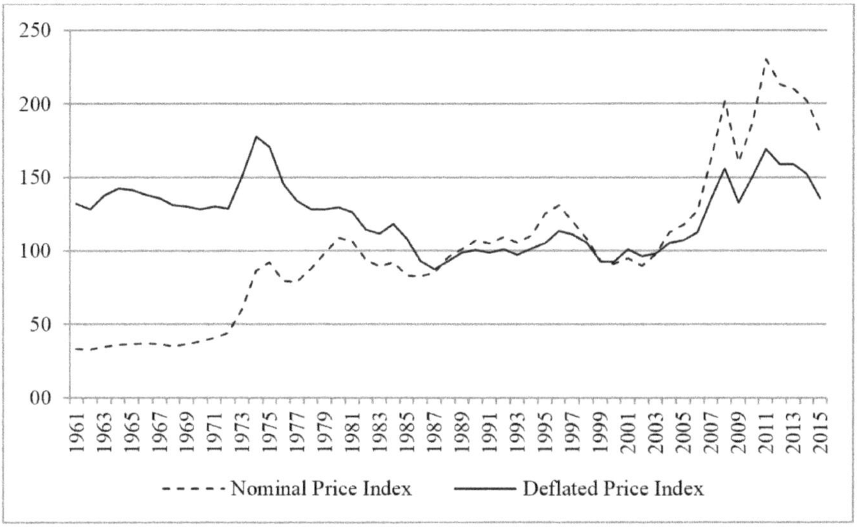

Figure 3.1 Global food price index between 1961 and 2015

Source: FAO (2016).

2012) and 67 million hectares (Zagema, 2011). Cotula (2012) reviewed the methodologies of estimating the scale of land acquisitions based on different sources and methods. Methodologies based on media reports often overstate the scale of the deals. Cotula (2012) noted that the 1,000-hectare threshold appears to filter out a large number of deals since majority individual deal sizes are less than 1,000 hectares. In addition to the methodological issues, implementation periods of investment deals in developing countries take longer due to the various obstacles in commercial penetration. It is thus difficult to identify the exact size of the land utilized in land deals. Investments in land are mainly intended for agricultural production but also include land deal data for mining, timber and nature conservation.

Investors involved in recent large-scale land acquisitions (LSLAs) include multinational corporations, state-owned enterprises, state agencies, private investors and pension funds with diverse interests, perceptions and mechanisms. Many point out that LSLAs are not entirely a new phenomenon (Hall, 2011; White *et al.*, 2012; Goetz, 2015). Complex and multifaceted interests and drivers of investors are key to understanding recent land acquisitions compared to colonial land grabs. The transnational aspect of the new land acquisitions has been highly emphasized due to its relevance to the global value chains of agricultural commodities. Newly emerged investors sponsored by governments suggest the motivation in securing scarce resources such as land and water to grow food. For example, some of the state-owned enterprises linked to China appear to be interested in contributing to food security back home. Yet, the scale and role of state-sponsored investors are disputable. Cotula (2012) points out that the role of East Asian and the Gulf State governments are largely overstated in media reports.

It is important to note that while global databases have improved in the past few years, it is important to treat them with scepticism. Many have noted that there are cases of double counting and some input into the databases are based on media reporting of political statements and speculations, not least in relation to investments by East Asian governments and corporations linked to them (Jägerskog *et al.*, 2012; Keulertz, 2013; Woertz, 2013b). Often these databases are used without much critical analysis of the data, thus perpetuating the problems associated with the data in the first place. Another key aspect of the data challenge relates to the implementation of land deals. After deals are announced, some land investment projects have not been implemented on the ground. The process of installing project administration can be followed by the challenges on the ground. Furthermore, challenges when investments have been made, such as those relating to governance (e.g. protests from local communities) have often delayed or halted implementation (Keulertz and Woertz, 2015). Far too much of the early research on land acquisition unfortunately used the data in a manner that led to a general overestimation of:

1 the number of deals that had been signed;
2 the size of the area grabbed/acquired/leased; and
3 the level of the implementation of the deals (often the signing of a deal was seen as equal to a deal).

Quantitative data collection of land deals presents many difficulties, and the accuracy and reliability of databases are highly questionable (Oya, 2013). The fact that the initial quality of the data was relatively poor was, of course, both a reason that the area was seeing a significant expansion in the wake of the 2007–2008 price crisis but also that the research on the topic was in many aspects in its infancy. Scoones *et al.* (2013) argue that the methodology to get the facts right in land acquisition research is crucial and call for a second phase of research that employs more specific, grounded and transparent methods instead of the aggregated and hectare-focused methods. Such an approach would also lead to a more developed understanding of the context-specific circumstances of various deals and what effect they have (or may have) on issues such as land rights, water rights, pastoralist movement, water sharing, etc.

Blue and green water considerations

From a water management perspective, it can also be noted that the early research almost fully excluded water aspects (Jägerskog *et al.*, 2012; Woodhouse, 2012a). This is evident not only by looking at the research but also in relation to the international efforts of establishing accepted principles for how investors could/should address key aspects of land acquisitions. In the Responsible

Agricultural Investment principles developed under the lead of the Food and Agriculture Organization (FAO) and the World Bank, water-related aspects are not seriously addressed (Borras and Franco, 2014). Often the contracts and agreements signed on land are silent on water requirements. Water seems to be taken for granted in the deals and not taken into account. This is in many aspects a reflection of the lack of a collaborative approach between land and water management. For a proper understanding of the water-related aspects of the land investments, an approach is needed that take into account the various types of waters that are being used for the investments. Is it blue water from rivers, lakes or aquifers? Or is it green water?[4] Or is it perhaps treated wastewater (grey water) used for irrigation? This is important, since depending on what type of water that is being used will have different effects on the socio-economic and political aspects of the deals. At present there are no reliable macro-level data available that distinguish what type of water is being used in the various land investments. These questions endorse the necessity to disaggregate 'water' in investments in relation to context-specific situations. Improved data in water use planning could, for example, prevent conflicts caused by the investments.

Questions also need to be raised on the seasonal variation of water resource availability and its appropriateness for the intended crop production. It is important to understand this in order to have an understanding of the 'full' picture (Falkenmark *et al.*, 2014). As water is a renewable resource, it is important to understand that it is a flux and not a stock. The conceptualization of green and blue water is, to some degree, based on how water is used in food production. First, *green water* is the precipitation that is absolved by biomass such as forest grasslands, wetlands and crops which evaporates into the atmosphere. *Blue water* is the water that infiltrates into aquifers or flows to lakes, rivers and dams (see Figure 3.2). Green water is not as visible as blue water since it cannot be stored in a basin but is instead found in soil or biomass. It is thus not surprising that the focus of the international community has, primarily, been on blue water (Falkenmark and Rockström, 2006). However, green water is crucial for food production. At the global level, green water accounts for about 80 per cent of the total agricultural water use and the remaining 20 per cent comes from blue water sources (Molden, 2007). In the Nile Basin, a mixed picture is presented on the importance of blue water and green water in agricultural production. In Egypt, more than 75 per cent of agricultural production comes from irrigated areas; but in the rest of the basin countries, either more than half or 75 per cent of production comes from rain-fed areas (Molden, 2007). Many developing countries are heavily dependent on green water for their food production. In sub-Saharan Africa, for example, around 96 per cent of the farmland is rain-fed.[5] Woodhouse (2012a) has emphasized green water in land investments and highlighted the ambiguity related to it in terms of water requirements for agricultural projects.

Falkenmark *et al.* (2014) identify how land acquisitions can affect water in the host country of the investments. It will, first, lead to land-use changes by

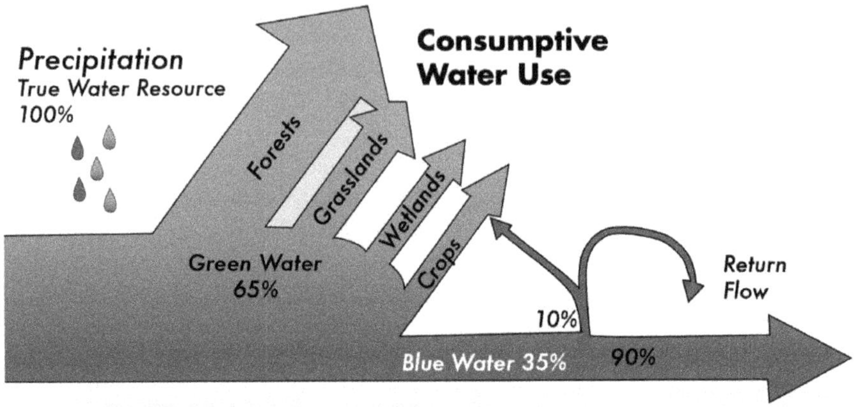

Figure 3.2 Blue and green water

Source: SIWI *et al.* (2005).

increasing water productivity and green-water use on croplands which will reduce stream flow; and, second, contribute to developing pasture areas into productive croplands that will turn some green water into blue water.

MENA countries challenging the virtual water hegemons?

The global grain market has 'an hourglass shape' where only a handful of trading companies deal with numerous farmers and consumers in the supply and demand side. The structural composition of the market creates an uneven power relationship between the powerful global grain trading companies and the rest. Allan (1997) coined the term 'virtual water' which analyses other forms of water such as water in food and other produce. Since water is so vital, controlling water resources in water-scarce regions has been an issue of contention. The power to control virtual water is also subject to contention. Sojamo *et al.* (2012) argue that some of the investments in agricultural production can be seen as a counter-hegemony strategy against 'virtual water hegemony' in the global grain market. As a result, the investments made by outside actors (states, corporations or other actors) will compete with the current rights-holders to gain control over the water. In order not to be solely dependent on the major private-sector actors who control much of the international food markets, investments can provide more control.

In the Nile context, the MENA countries and in particular the countries comprising the Gulf Cooperation Council (GCC)[6] have invested significantly in agriculture. Out of the seven top land investments abroad by the GCC, four of them are located in the Nile region (Sudan is number one, Ethiopia number four, Egypt number five and Tanzania number seven). This shows the strong MENA interest in the Nile region (Woertz and Keulertz, 2015).

A further potentially contentious aspect is whether the required water resource is sourced from a transboundary basin, thus bringing in added complexity and potential for conflict (Jägerskog *et al.*, 2012). States are, thus far, the principal actors in the international political processes around water sharing, but influence by outside actors can be also factored into the political processes. Negotiations on transboundary waters are done between states. Still, outside actors who are investing in a country become relevant in the hydropolitical landscape in a river basin such as the Nile Basin. The powerful outside actors (states outside of the basin, private sector actors, etc.) can coerce local authorities and communities if their interests are not being protected, such as in the case when a sufficient amount of acceptable quality water is not provided for their agricultural investments.

The fact that water routinely is ignored as a factor makes the situation worrisome since an integrated perspective between water and land management is needed in order to understand if a particular investment has the consent by all actors in a given geography. The ubiquitous consents by all actors in the regional, national and local levels are hard to achieve on any deals. Falkenmark *et al.* (2014) even suggest that the concept of integrated water resource management (IWRM) should have an added 'L' for 'Land' in order to integrate all perspectives. By including land-use aspects in water management, more of the trade-offs and the (sometimes) incompatibility between proposed land use and water availability will become apparent.

Global trends in land acquisitions: drivers and constraints

Acquiring agriculture products from overseas such as tea, coffee, spices and other cash crops is not a new phenomenon. It has been one of the main drivers of global trade since the fifteenth century (Wallerstein, 2011). During the colonial periods, plantations and contract farming were practised, and irrigation infrastructure was developed overseas for agricultural production. Recent LSLAs share both similarities and differences with the earlier colonial agricultural expansion. Compared to the colonial era, the present international trade regime is more integrated with consumer markets and producers all around the world through the enlarged role of transnational corporations.

Many have researched the recent phenomenon of LSLAs in relations to the scale, scope of contracts, governance and rights aspects (Deininger and Byerlee, 2011; Matondi *et al.*, 2011; Zagema, 2011; Anseeuw *et al.*, 2012; Cotula, 2012; Havnevik, 2014). Torres (2012) critically addressed distributional justice and the role of economic structure in land acquisitions. The linkages between the international agricultural commodity trading regime and foreign investment dynamics have been researched related to land acquisitions (Peterson and Garland, 2010; Sojamo *et al.*, 2012; Toft, 2013). Clapp (2014) points to the unreliable international food trade regime as a factor contributing to the increased interest to invest in farmland followed by volatile agricultural commodity prices and raised input prices. The actors of recent land acquisitions are extremely diverse in their ownership structures and operational models. The diversity of actors and their modalities reflect the current stage of the international trade regime that allow actors to penetrate national economic barriers (Cohen, 2007).

The regulative role of the host country government still remains significant as an enabling factor especially in land acquisitions (Cuffaro and Hallam, 2011). However, if a proper and reliable governance system is not in place in a country, this may also work contrary to the interest of the investor in the sense that there is less security for the investment. Investment treaties such as Bilateral Investment Treaties (BITs) and International Investment Agreements serve the purpose of protecting investors' interests by implementing legal instruments such as preferential treatment, non-discrimination, guarantee against expropriation and nationalization, and compensation for losses. The proliferation of BITs plays a role in facilitating the influx of foreign direct investments (FDIs) on farmland (Peterson and Garland, 2010). Countries in the Nile Basin have signed 231 BITs since the 1970s until recently.[7] Most BITs have been signed in Egypt, followed by Sudan and Ethiopia. The signing parties of BITs often include investor countries from North America, Europe, East Asia, as well as the MENA region. In addition to the BIT, multilateral international investment treaties are in force in the Nile Basin countries. Multilateral investment treaties such as the Agreement on the Promotion, Protection and Guarantee of Investments amongst the Member States of the Organization of the Islamic Conference aim to extend the cultural and religious ties to the economic realm. The agreement is constituted with 56 states and one territory, including Egypt and Sudan as parties, and aims to facilitate investments between parties including the GCC investors and some countries in the Nile Basin.

Water is both a driver and a constraint of land deals (cf. Beyene and Sandström, 2016). Recent LSLAs are, however, not reflected in water-sharing agreements and water allocation processes (Jägerskog *et al.*, 2012; Woodhouse, 2012a). Woodhouse (2012b) points out the lack of consideration of how investments in land may affect, directly or indirectly, the water situation in a basin. The availability of water is one of the constraints of the successful implementation of land investments, and the challenge becomes more pronounced when the investor needs to secure additional water resources in order to produce intended crops. Investors may need to construct new irrigation facilities or rehabilitate existing irrigation facilities to gain access to additional 'blue water'. Unlike 'green water', rainwater consumed by crops, infiltrated in soil and evaporated to the atmosphere, access to blue water is crucial to enable intensive crop production in regions where precipitation is varied and insufficient.

Large-scale land acquisitions have been seen as a potential driver for a significant increase in virtual water trades between water and land resource-endowed countries and countries with inadequate land and water resources. Rulli *et al.* (2013) have estimated the green water required for implementing concluded land deals worldwide, which currently account for $379 \times 10^9 \, m^3 \, yr^{-1}$, an amount sufficient to produce enough food to feed about 300–390 million people. This estimation is based on the data compiled from a number of sources in May 2012 and summed up to 46.94 million hectares.

Land acquisitions as a source of water conflict

Reallocation of water resources to accommodate new land deals often causes tension and conflicts between various water users (Houdret, 2012). In addition to the insufficient water-specific provisions on contracts, the challenge in utilizing water for land investments comes from the 'fluid' nature of water flowing through regional and national boundaries and continuously changing its forms. If the water used for the plantations of the investments is largely rain-fed ('green water'), land deals are likely to be less conflict-inducing than if it is blue water that is shared by different regions or countries. Annual variability of rainfall also contributes to the difficulties to quantify the amount of water consumed in agricultural projects. Falkenmark *et al.* (2014) note that there are trade-offs between green and blue water systems within the same basin. If, for example, green water is mainly used for agriculture, the water may not percolate into groundwater systems but instead produce run-off and 'become' blue water. On the transboundary level, the water situation in the basin can change through the loss of blue water downstream as a result of the consumptive use of green water upstream. On the flipside of land investment being a source of conflict in the regions where land is acquired, it may also work as an ameliorating factor in the countries that are making the investment. The MENA investment in the Nile Basin may function as a means to decrease water conflict (within and across boundaries) in the water-scarce MENA region as it eases off some of the pressure on their water resources (Jägerskog and Kim, 2016). This requires, however, an increased implementation on the part of the MENA countries in the Nile Basin, where implementation has been lagging behind (Woertz and Keulertz, 2015). At the same time, while an increased implementation of the MENA-related investments in the Nile Basin could improve food security in the investing countries, it will at the same time negatively affect the food security in the Nile Basin as water will be 'diverted' from the Nile Basin system.

Food security as a national security concern

Ensuring domestic food supply through overseas investments has become a strategic policy on a national level in countries lacking the arable land to produce enough food, including Gulf States and East Asian countries. Food crops and grains such as rice, wheat, cassava, beans and corn are common in agricultural land investments. Traditional cash crops and water-intensive vegetables are also popular in large-scale land deals where water supply is considered sufficient.

The MENA region and, in particular, the Gulf States have increased their investments in agriculture in foreign lands after the food price crisis of 2007–2008, not least in the Nile Basin. The overall ratio in the Arab region of food self-sufficiency in 2008 was around 60 per cent and with increasing population, potential decreased rainfall, affected by climate change, that figure is estimated to decrease to around 35 per cent by 2050 (Elhadj, 2008). During the 1970s, Arab land investments in the Nile Basin were primarily targeting Sudan but have successively shifted to include other countries in the basin, primarily Ethiopia, Egypt and Tanzania (Woertz and Keulertz, 2015; Swain and Jägerskog, 2016). It was not necessarily the food price spikes that created the surge in looking for land investments opportunities in the region but rather the export restriction of key commodities (which also contributed further to the price spikes) like wheat from countries such as Russia and other central wheat exporters (Woertz, 2013a, 2013b; Swain and Jägerskog, 2016). The MENA region has responded to the food price crises and food exporting restrictions by increasing outgoing investments in agriculture production abroad. Countries with religious and cultural ties with the MENA region, such as Sudan, Egypt, Ethiopia and Tanzania but also Kenya, are preferred as investment destinations (Woertz and Keulertz, 2015; see also Chapter 2 in this volume).

Global trends in land acquisitions: investors water scarce and recipients water abundant

One of the new aspects of recent LSLAs is the diversity of actors. Non-state actors and state-owned enterprises have emerged as some of the important investors. Investors of recent land acquisitions come from Europe and North America as well as newly developed or emerging economies, including China, South Korea and the Gulf States. New actors from East Asia and the Gulf region have been experiencing shortages in water and arable farmland for food production, which has meant that domestic agricultural expansion is no longer possible for them. In the MENA context, this is evident in the wealthier Gulf countries where a more meat-intensive diet is observed (Woertz and Keulertz, 2015). This trend might boost the motivation to acquire land overseas by countries expecting increasing food imports.

It is not a coincidence that most of the land acquisitions take place in regions with plentiful water resources. However, some of the basins where the investments occur are not always well sourced with water. Taking into account the predicted population growth in the countries of the Nile Basin, there will be increasing water scarcity in some areas. This makes the investments (and the country where the investments have been made) much more vulnerable and potentially politically charged.

Land and water acquisitions in the Nile Basin

The Nile Basin is a popular destination for land acquisitions and investments. Most of the investments occurred in the aftermath of the 2007–2008 food price crisis through a range of investors. The land and water connections in the region (as elsewhere) have been poorly documented (or sometimes absent)

(Cotula, 2009; Woodhouse, 2012a). Investors have been from a variety of geographies and have had different interests and goals. Some have invested from a private-sector perspective, others for speculation, and yet others for food security (in the investing countries). Inadequate attention has been paid to the water aspects of the deals, often not even including water as a factor; when included, there has been ignorance of *what* type of water should be used for the investments. Unless secure access to water is present, the investment risk is high. If the land investment is dependent on rainfall, the challenges being brought by climate variability and change need to be taken into account. A pertinent question is if the water users compete with other either existing or potential water users in the country or even with neighbouring countries, does this make the issue of land investments a regional, hydropolitical matter?

The hydropolitics of the Nile Basin are complex. The last decade(s) have seen increased cooperation in the basin through the Nile Basin Initiative (NBI), which has focused on development programmes in the basin. Parallel to the more development-oriented programmes of the NBI, negotiations have taken place on a new Cooperative Framework Agreement (CFA) to form an overarching agreement for the whole Nile Basin, something that has been previously lacking. Previous agreements from 1929 and 1959 are non-inclusive and cover, in principle, Egypt and Sudan. While those agreements are not seen as being enough (or even relevant) by the upstream Nile Basin states, Egypt and Sudan are keen on maintaining the key features of those agreements which secured them a large part of the flow. In particular, this is considered very important for Egypt. After lengthy negotiations, the CFA was signed by a number of the riparian countries in 2011, against the wishes of Egypt and Sudan. For a more detailed account of the political process, see Chapter 4 in this volume. It is clear that land acquisitions/investments have not featured as a key component of water-related agreements (not the old ones from 1929 and 1959 and not the CFA). It is, however, noted – in this volume as well as by Jägerskog *et al.* (2012) – that the land acquisitions/investments aspects ought to be considered from a transboundary and hydropolitical perspective as the investments in the basin are bound to either directly (through the use of Nile waters) or indirectly (through the use of green water/rainwater that may otherwise have produced run-off or formed groundwater).

MENA–Nile investment connection

The MENA region has many complex and interlinked dynamics: intensive and long-lasting armed conflicts that worsen water scarcities (Jägerskog and Kim, 2016). Historically, there have been different 'waves' of interest in the Nile region from the MENA countries (Swain and Jägerskog, 2016). During the 1970s, much pressure came from the League of Arab States (LAS) on Sudan, which was acting as a breadbasket for the region as it had a relative abundance of both water and arable land. The close proximity to the Nile region is also relevant, both in a historic but also contemporary perspective, not least against the backdrop of the conflict between largely the Arab States (Egypt and Sudan) and the non-Arab States in the Nile context. Finally, when looking at MENA and Nile Basin land investment, it also illustrates that the issue of land investment is easily 'securitized' when water, land and food securities and scarcities are linked together (Swain and Jägerskog, 2016).

The Arab Organization for Agricultural Development (AOAD) was established to coordinate agricultural expansion and devise integration of agriculture in the region. However, due to poor governance and lack of capacity, among other things, the idea of Sudan as a breadbasket for the region never really materialized (Keulertz, 2013). Yields per hectare as well as per unit of water used in Sudan are very low today, both from a regional perspective and even more in an international comparison (Woertz, 2013a). Instead, the region became reliant on the import of virtual water to counter its food (and water) needs. In the wake of the food price crisis, primarily the Gulf States but also countries such as Jordan have increased their investments abroad. Investments work as a means to increase food security and also to access water more reliably and not be fully dependent on virtual water flows in the global market. These 'flows' are largely beyond the control of international corporations and also sensitive to export bans that major wheat-producing countries employed (e.g. Russia) in the 2008 food price crisis.

Jägerskog and Kim (2016) have analysed the idea that some countries, in an effort to further gain control over their food security, invest in agricultural land in the Nile Basin (and other parts of the world as well) and thereby create their 'own' virtual water flows. It is also fair to assume that the dependency on the Nile Basin may increase for MENA countries as investments by Saudi Arabia, the United Arab Emirates (UAE) and Qatar in the region are expanding. While the historic narrative in most, if not all, MENA countries has been on food self-sufficiency, the realization that both water scarcity as well as lack of arable land present major challenges. Saudi Arabia has relatively recently given up its grand project of growing wheat in the desert (which at some point made them the world's fifth largest wheat exporter) and adopted a policy of relying on import. Therefore, increasing control through own investments in foreign countries also carries a security perspective for many of the countries in the MENA region (Swain and Jägerskog, 2016).

It is clear that there are many reasons for MENA investors to look to the Nile Basin for investments in land. First, the experiences of the recent food price crises of 2007–2008 and 2011–2012, exacerbated by, for example, Russian export bans on wheat, were felt in the Arab region where prices of staple commodities such as wheat increased, drastically affecting social and political stability (Swain and Jägerskog, 2016). This necessitated a need to look for investments abroad as water scarcity in the region continued to increase as MENA populations grow, and climate variability (often meaning less water) continued to be a factor. Second, the cultural ties between the MENA region and the Nile Basin region are strong, particularly in Sudan. Already in 1970 the LAS aimed to make Sudan a 'breadbasket' of the region, although with little success. Thus, it makes sense for water-scarce countries in the MENA region to look to the Nile region to ensure food and water security. Third, the abundance of capital available for investments in some Arab countries (through, for example, Sovereign Wealth Funds) also provided an opportunity to enter the Nile basin. Fourth, policy changes in the Arab region, most significantly Saudi Arabia, of abandoning its massive domestic cereal production programme have led to increased interests in investments in other regions such as the Nile Basin. Fifth, the Nile Basin also has close proximity to the MENA region and can thus be seen to be a good investment since it is 'close to home'.

To use the arguments presented by Sojamo *et al.* (2012), it seems as if the countries in the Arab region are using land investments in the Nile Basin (and beyond) as a counter-hegemonic strategy in order to counter the hegemony of the major international food corporations and therefore somewhat decrease reliance on the global market for food. By controlling, to some extent at least, the land used for food production, food security (and control over it) is protected.

Conclusions

This chapter has focused on the investments from MENA in the Nile Basin region. The analysis of MENA investments has been done against the backdrop of the current trends in the research on global land acquisitions from the quantitative tradition driven by global-scale databases. The chapter especially focuses on the conceptual and empirical link between food security and water scarcity, which is accentuated in the case of MENA. Despite the conjuncture between two urgent and important policy priorities (food and water security), water-related aspects are routinely overlooked in the agricultural investment deals and in the domestic and international policies in foreign investments. The chapter has also highlighted the need to go beyond the general analysis of *how much* water is being used in the land investment schemes to also look at *what type* of water is being used. For example, is it national or transboundary? Is it blue water or green water? Such considerations are important if one wants to understand the socio-political aspects of the investments in the Nile Basin. Whether it is blue water or green water makes a substantial difference, not least from a transboundary perspective.

A number of conclusions have been highlighted in this chapter regarding the importance for the MENA–Nile Basin investment link. Changes in the MENA region, in terms of decreasing availability of water on a per capita basis, and increased pressure on the already scarce water resources in the region brought about by climate change, are some of the reasons why it has become a priority for MENA countries to search for alternative solutions to achieve food security.

Furthermore, close cultural ties between the Arab region and at least part (primarily Egypt and Sudan) of the Nile Basin, as well as geographical proximity, has arguably lead to increased interest from MENA investors. It remains, however, to be seen how many of the investment deals that have been signed by MENA investors in the Nile Basin region will materialize and what political effects this will have in the MENA 'home' region. Will political relations between the investing countries and the hosting countries become stronger? Will some of the Gulf countries become indirect hydropolitical actors in the Nile Basin? In addition, how will the investments and their consequences play out in the Nile Basin itself? Will it lead to improved relations in the basin? Or will it lead to increased competition in the region?

Notes

1 UN DESA, Population Division, *The 2015 Revision* of *World Population Prospects*.
2 See www.un-foodsecurity.org/background (accessed 30 March 2016).
3 Land Matrix: www.landmatrix.org (accessed 26 April 2015).
4 Green water refers to the water in soil moisture and water in biomass that is critical for agricultural production.
5 FAO (2011). *The State of the World's Land and Water Resources for Food and Agriculture (SOLAW): Managing Systems at Risk*.
6 The GCC consists of Kuwait, Saudi Arabia, Bahrain, Qatar, the UAE and Oman.
7 UNTAD database: http://investmentpolicyhub.unctad.org/IIA (accessed 5 January 2016).

References

Allan, J. A. (ed.). (1997). *Virtual Water: Tackling the Threat to our Planet's Most Precious Resource*. London: I.B. Tauris.

Anseeuw, W., Boche, M., Breu, T., Giger, M., Lay, J., Messerli, P. and Nolte, K. (2012). *Transnational Land Deals for Agriculture in the Global South*. The Land Matrix Partnership. Retrieved from http://landportal.info/landmatrix/media/img/analyticalreport.pdf (accessed 20 January 2016).

Beyene, A. and Sandström, E. (2016). Emerging water frontiers: Large-scale land and water acquisitions and food security in Africa. In Tvedt, T. and Oestigard, T. (eds) *Water and Food: Africa in a Global Context*. London/New York: I.B. Tauris.

Borras, S. and Franco, J. (2014). From threat to opportunity? Problems with the idea of a 'code of conduct' for land-grabbing. *Yale Human Rights and Development Journal*, *13*(2), 7.

Brown, L. R. and Black, E. R. (1970). *Seeds of Change: The Green Revolution and Development in the 1970s*. New York: Praeger.

Clapp, J. (2014). Financialization, distance and global food politics. *Journal of Peasant Studies*, *41*(5), 797–814.

Cohen, S. D. (2007). Multinational corporations and foreign direct investment: Avoiding simplicity, embracing complexity. *OUP Catalogue*. Oxford: Oxford University Press.

Conway, G. and Wilson, K. (2012). *One Billion Hungry: Can We Feed the World?* Ithaca, NY: Comstock Pub. Associates.

Cotula, L. (2009). *Land Grab or Development Opportunity? Agricultural Investment and International Land Deals in Africa*. London: IIED.

Cotula, L. (2012). The international political economy of the global land rush: A critical appraisal of trends, scale, geography and drivers. *The Journal of Peasant Studies*, *39*(3), 649.

Cuffaro, N. and Hallam, D. (2011). 'Land grabbing' in developing countries: Foreign investors, regulation and codes of conduct. *Regulation and Codes of Conduct* (20 January).

Deininger, K. W. and Byerlee, D. (2011). *Rising Global Interest in Farmland: Can it Yield Sustainable and Equitable Benefits?* Washington, DC: World Bank.

Earle, A., Cascão, A. E., Hansson, S., Jägerskog, A., Swain, A. and Öjendal, J. (2015). *Transboundary Water Management and the Climate Change Debate*. London: Routledge.

Elhadj, E. (2008). Dry aquifers in Arab countries and the looming food crisis. *Middle East Review of International Affairs*, *12*(4), 1–11.

Falkenmark, M. and Rockström, J. (2006). The new blue and green water paradigm: Breaking new ground for water resources planning and management. *Journal of Water Resources Planning and Management*, *132*(3). 129–132.

Falkenmark, M., Jägerskog, A. and Schneider, K. (2014). Overcoming the land–water disconnect in water-scarce regions: Time for IWRM to go contemporary. *International Journal of Water Resources Development*, *30*(3), 391–408.

FAO (2016). FAO Food Price Index in nominal and real terms. Retrieved from www.fao.org/worldfoodsituation/foodpricesindex/en (accessed 27 June 2016).

Goetz, A. (2015). How different are the UK and China? Investor countries in comparative perspective. *Canadian Journal of Development Studies/Revue Canadienne d'études du développement*, *36*(2), 179–195.

Hall, R. (2011). Land grabbing in Southern Africa: The many faces of the investor rush. *Review of African Political Economy*, *38*(128), 193–214.

Havnevik, K. (2014). *Responsible Agricultural Investments in Developing Countries: How to Make Principles and Guidelines Effective*. Stockholm: Swedish FAO Committee.

Houdret, A. (2012). The water connection: Irrigation, water grabbing and politics in southern Morocco. *Water Alternatives*, *5*(2), 284–303.

IFPRI. (2015). *2015 Global Hunger Index*. International Food Policy Research Institute. Retrieved from http://ghi.ifpri.org (accessed 20 January 2016).

Jägerskog, A. and Kim, K. (2016). Land acquisition: A means to mitigate water scarcity and reduce conflict? *Hydrological Sciences Journal*, *61*(7), 1338–1345.

Jägerskog, A., Cascão, A., Håsmar, M. and Kim, K. (2012). *Land Acquisitions: How Will They Impact Transboundary Waters* (SIWI Report No. 30). Stockholm: SIWI.

Jiménez Cisneros, B. E., Oki, T., Arnell, N. W., Benito, G., Cogley, J. G., Döll, P. and Mwakalila, S. S. (2014). Freshwater resources. In C. B. Field, V. R. Barros, D. J. Dokken, K. J. Mach, M. D. Mastrandrea, T. E. Bilir, … L. L. White (eds), *Climate Change 2014: Impacts, Adaptation and Vulnerability. Part A: Global and Sectoral Aspects. Contribution of Working Group II to the Fifth Assessment Report of the Intergovernmental Panel on Climate Change* (pp229–269). Cambridge and New York: Cambridge University Press.

Keulertz, M. (2013). *Drivers and Impacts of Farmland Investment in Sudan: Water and the Range of Choice in Jordan and Qatar* (PhD thesis). London: King's College London.

Keulertz, M. and Woertz, E. (2015). States as actors in international agro-investments. *International Development Policy: Revue Internationale de Politique de Développement*, *6*(1).

Lundqvist, J., Grönwall, J. and Jägerskog, A. (2015). *Water, Food Security and Human Dignity: A Nutrition Perspective* (Publication series No. 10). Stockholm: Swedish FAO Committee.

Matondi, P., Havnevik, K. and Beyene, A (eds) (2011). *Biofuels, Land Grabbing and Food Security in Africa*. London: Zed Books.

Molden, D. (2007). *Water for Food Water for Life: A Comprehensive Assessment of Water Management in Agriculture*. London: Earthscan.

Oya, C. (2013). Methodological reflections on 'land grab' databases and the 'land grab' literature 'rush'. *Journal of Peasant Studies*, *40*(3), 503–520.

Peterson, L. E. and Garland, R. (2010). *Bilateral Investment Treaties and Land Reform in Southern Africa* (*Droits et Démocratie*). Centre international des droits de la personne et du développement démocratique. Retrieved from www.peacepalacelibrary.nl/ebooks/files/334103282.pdf (accessed 20 January 2016).

Porter, J. R., Xie, L., Challinor, A. J., Cochrane, K., Howden, S. M., Iqbal, M. M., … Travasso, M. I. (2014). Food security and food production systems. In C. B. Field, V. R. Barros, D. J. Dokken, K. J. Mach, M. D. Mastrandrea, T. E. Bilir, … L. L. White (eds), *Climate Change 2014: Impacts, Adaptation and Vulnerability. Part A: Global and Sectoral Aspects. Contribution of Working Group II to the Fifth Assessment Report of the Intergovernmental Panel on Climate Change* (pp485–533). Cambridge and New York: Cambridge University Press.

Rulli, M. C., Rulli, M.C. and D'Odorico, P. (2013). The water footprint of land grabbing. *Geophysical Research Letters, 40*(23), 6130–6135.

Scoones, I., Hall, R., Borras Jr, S. M., White, B. and Wolford, W. (2013). The politics of evidence: Methodologies for understanding the global land rush. *Journal of Peasant Studies, 40*(3), 469–483.

SIWI, IFPRI, IUCN and IWMI (2005). *Let it Reign: The New Water Paradigm for Global Food Security*. Final Report to CSD-13. Stockholm: Stockholm International Water Institute.

Sojamo, S., Allan, J., Keulertz, M. and Warner, J. (2012). Virtual water hegemony: The role of agribusiness in global water governance. *Water International, 37*(2), 169.

Swain, A. and Jägerskog, A. (2016). *Emerging Security Threats in the Middle East: The Impact of Climate Change and Globalization*. Lanham, MD: Rowman & Littlefield Publishers.

Toft, K. H. (2013). Are Land Deals Unethical? The Ethics of Large-Scale Land Acquisitions in Developing Countries. *Journal of Agricultural and Environmental Ethics, 26*(6), 1181–1198.

Torres, I. V. (2012). Water grabbing in the Cauca basin: The capitalist exploitation of water and dispossession of afro-descendant communities. *Water Alternatives, 5*(2), 431–449.

UN (2015). *The 2015 Revision of World Population Prospects*. New York: DESA, Population Division.

Wallerstein, I. (2011). *The Modern World-System I: Capitalist Agriculture and the Origins of the European World-Economy in the Sixteenth Century, with a New Prologue* (Vol. 1). Berkeley, CA: University of California Press.

White, B., Borras, S. M., Hall, R., Scoones, I. and Wolford, W. (2012). The new enclosures: Critical perspectives on corporate land deals. *Journal of Peasant Studies, 39*(3-4), 619–647.

Woertz, E. (2013a). *Oil for Food: The Global Food Crisis and the Middle East*. Oxford: Oxford University Press.

Woertz, E. (2013b). The governance of Gulf agro-investments. *Globalizations, 10*(1), 87–104.

Woodhouse, P. (2012a). Foreign agricultural land acquisition and the visibility of water resource impacts in sub-Saharan Africa. *Water Alternatives, 5*(2), 208–222.

Woodhouse, P. (2012b). New investment, old challenges: Land deals and the water constraint in African agriculture. *Journal of Peasant Studies, 39*(3–4), 777.

Zagema, B. (2011). Land and power: The growing scandal surrounding the new wave of investments in land. *Oxfam Policy and Practice: Agriculture, Food and Land, 11*(6), 114–164.

Zeitoun, M. and Warner, J. (2006). Hydro-hegemony: A framework for analysis of trans-boundary water conflicts. *Water Policy, 8*(5), 435–460.

4 The Gulf states in the political economy of the Nile Basin

A historical overview

Harry Verhoeven

Introduction

Traditional accounts of Nile Basin politics tend to focus on the trajectories of the region's key states – Egypt, Ethiopia, Sudan – and the international legal context of disputes over water. For many, the central dynamic has been the classic conflict-cooperation dichotomy (Wiebe, 2001; Swain, 2002; Mekonnen, 2010), which has led to attempts to explain what conditions need to be met to transition from adversarial relations to more collaborative ones (Howell and Allan, 1994; Waterbury, 1997; Wu and Whittington, 2006; Salman, 2013). Those working in this vein have mostly concentrated, following the title of Waterbury's classic contribution, on the national determinants of regional hydropolitics (Waterbury, 2002; Salman, 2011), with a supportive role for Western-dominated international financial institutions such as the World Bank, whose capital has often been a necessary, if not sufficient, condition for transforming basin politics around the world. As the West has faced relative decline in the past ten years, scholarship has increasingly dissected the impact of emerging financiers of water resources development projects such as China (Cascão, 2009; Verhoeven, 2011a, 2011b).

Concentrating on the Nile Basin states, Western powers (Britain as colonial hegemon and later the USA as global superpower) and more recently China has obscured the important longitudinal role played by Gulf states in regional relations generally and hydropolitics in particular. The mass provision of oil and gas to international markets is but one dimension of the Gulf's rise to global prominence (Ulrichsen, 2011). The leading polities – Saudi Arabia, Kuwait, the United Arab Emirates (UAE) and Qatar – have long seen Africa's north-eastern corner as a logical extension of their socio-cultural, economic and political sphere and have, through exchanges characterized by both animosity and partnership, influenced the fortunes of states and regimes there. This historical proximity has in the last decades very much been an asymmetric relationship with people and produce moving to the Arabian Peninsula from the basin, and capital and ideological currents flowing the other way across the Red Sea: an inversion of older trends. In a context of enduring statehood but simultaneous state fragility in the Nile Basin, this asymmetry has had profound consequences in that both domestic political economies and intra-basin relations have been shaped by Gulf Arab interests in important ways.

The Nile has for decades been not just the main ecological highway of Northeast Africa but also its most important political artery, with elites trying to harness its economic potential for state-building (Verhoeven, 2015a, 2015b, 2015c). The idea that control over water equals control over people and that, hence, water is power has proven deeply consequential: the question of how the river's waters can be put at the service of internal power consolidation and economic expansion has throughout the modern period been a key determinant of the behaviour of ruling elites in their external policy. This has given rise to a zero-sum paradigm in the basin in which the relative gains of one party are seen to imply a material degradation and political humiliation for the others. In this unforgiving strategic environment, the fortunes of incumbents are heavily influenced by their ability to leverage outside resources for internal battles. While Ethiopian relations with the Gulf states have been antagonistic for most of the past 60 years almost regardless of the ideological disposition of governments in Addis Ababa, regimes in Egypt and Sudan have intermittently been isolated from the Arabian Peninsula, only to reverse the situation a few years later and to draw on Gulf Arab financial resources and politico-military backing to secure domestic hegemony and expand their regional influence. In recent decades, big hydro-agricultural initiatives in Sudan and Egypt have not just depended on Gulf funding but the very ability of secular, military and Islamist regimes to remain hegemonic in their societies and the region has been moulded by relations with Riyadh and, more recently, Abu Dhabi and Doha.

This chapter is theoretically informed by the historical and international sociology approach to understanding states, their constituent social forces and their interrelations (Halliday, 2005). It opens with a brief sketch of the *longue durée* of relations between the Arabian Peninsula and the Nile Basin: after situating centuries of historical proximity, the focus moves to Egyptian 'hydro-hegemony', a term used to explore the subtle and not so subtle mechanisms underpinning a state's dominance in basin politics (Zeitoun and Warner, 2006). Gamal Abdel Nasser's bid for a revitalized Egypt via the construction of the Aswan High Dam and the 1959 Nile Waters Agreement (NWA) was initially supported by the Gulf Arabs. However, Cairo and Riyadh soon battled for leadership of the Arab world – a collision that ushered in a growing asymmetry between a stagnating Egypt and Gulf states bolstered by soaring oil prices. Following rapprochement between Sadat and the Al-Sauds, a different relationship emerged in the 1970s during which the Gulf became the Egyptian economy's financier, funded hydro-agricultural state-building in Ja'afar Nimeiri's Sudan and encouraged the containment of Marxist-Leninist and 'Christian' Ethiopia. Cairo's hydro-hegemony was increasingly underwritten by petrodollars, which also paid for Sudan's 'Breadbasket' vision to address exacerbating food security problems on the Arabian Peninsula.

The latter parts of this chapter outline the limits of the rising prominence of Gulf Arabs and their ability to pursue political-economic objectives and reshape the Nile Basin from the 1980s until today. Not only have the Gulf states increasingly vied for regional influence at each other's expense, they have regularly been frustrated in the pursuit of their interests, politically as well as economically. The failure of the Breadbasket gamble and the Agricultural Revival Programme in Sudan, as well as the rise to power of the military-Islamist Al-Ingaz regime in Khartoum, have generated security, commercial and political woes for Saudi Arabia, Kuwait and the UAE. It also contributed to the erosion of the hegemony of Egypt – Riyadh's most loyal ally – in the basin, to the advantage of Ethiopia, an old rival long distrusted on the Arabian Peninsula. This is the other side of the coin of influence: growing exposure and involvement also imply growing disappointments.

The antecedents to and origins of Egyptian hydro-hegemony

Ties between the Nile Basin and the Arabian Peninsula are ancient and today's relationships continue to be heavily mediated by historical experiences and the organic connections between peoples, markets and states woven over centuries. The basin has long loomed large in the imagination of Jews, Christians and Muslims alike. This religious dimension – and all the associated migratory flows and other exchanges it has evoked – remains a pivotal subtext to the sense of familiarity with which elites on both sides of the Red Sea approach one another. It was in Egypt that Moses confronted the Pharaoh and that the Holy Family sought refuge; the legendary kingdom of Kush, in today's Northern Sudan, features prominently in the Bible as the land ruled over by Noah's grandson; the first Muslim *hijra* to flee religious persecution in Mecca was to Ethiopia; and the mystical union of the Jewish Prophet-King Solomon and the Ethiopian Sheba remains an enigmatic story integral to Jewish, Christian and Islamic traditions. While none of these traditions venerate the Nile itself as a deity, the miracle of its flood and its life-giving properties have provided a natural décor for some of monotheism's most extraordinary narrations.

For centuries, no polity was dominant enough to bring both the Nile Basin and the Arabian Peninsula under one imperial hegemony – neither the Pharaohs nor the Roman Empire nor the Rashidun Caliphate. Successor Islamic empires such as the Fatimids, Abbasids and Mamelukes did manage to hold Egypt and Hejaz but struggled to push further south into Africa. It was not until the nineteenth-century reign of Muhammad Ali, a vassal of the Ottoman Empire who lorded over Cairo and enjoyed a huge degree of autonomy, that a ruler felt confident to penetrate deep along coastal Arabia and southwards into the Nile Basin. Ali consolidated the Hejaz and destroyed the first state of the Al-Sauds in 1818, reclaiming Mecca and Medina (Bowen, 2014: 90–96); he invaded Sudan in 1821.

Ali's foreign conquests happened, not coincidentally, in parallel to an audacious state-building project at home that aimed to revolutionize agricultural productivity and centralise decision-making over water, land and other resources; the ultimate objective was to turn the Pasha into a modern sovereign on the European model (Fahmy, 2002). Egyptian imperialism, rooted in elite competition in Cairo (Lawson, 1992), endured as subsequent Alawiyya Dynasty rulers pushed further towards the headwaters of the Nile, reaching Equatoria and the foothills of the Ethiopian Highlands and occupying swathes of modern-day Eritrea and Eastern Ethiopia. Throughout the nineteenth century, political and economic objectives entwined: the thirst for fertile land, slaves and control over water catalysed the formation of the modern Egyptian and Sudanese states. As I documented elsewhere (Verhoeven, 2015c), the idea that centralized control over the Nile and large-scale irrigated agriculture formed the bedrock of domestic and international power, became an ideological and material cornerstone of the emerging states and was internalized by the elites that dominated their bureaucracies, armed forces and agrarian economies.

Alawiyya imperialism ran out of steam about 50 years after Ali's invasions. The dazzling festivities to inaugurate the Suez Canal, the Ethio-Egyptian wars, the development of Cairo and Alexandria, growing contradictions in agriculture and resistance in the imperial peripheries such as Sudan all contributed to an escalating debt crisis for Egypt, which struggled to bear the twin cost of empire and state-building. The state that had sought to emancipate itself from Ottoman suzerainty and become an equal to the European powers found itself increasingly ensnared in Western imperialism. Cairo lost its independence to the notorious Caisse de la Dette – which ended Egyptian fiscal autonomy – and to Consul-General Evelyn Baring, who completed the country's integration into the British Empire (Mitchell, 1988). However, the crumbling of Alawi regional hegemony did not lead to a fracturing of the territorial control over the Nile Basin and Western Arabia that Muhammad Ali had established. While the Hejaz had passed back into the hands of the Ottoman Sultan after 1840, Egypt's anchoring in the British imperial orbit had knock-on effects for the wider region: because the Suez Canal was the vital military, commercial and logistical link between the metropolis and British India, the stability of Egypt became an imperial top priority. This meant both deep involvement in the domestic economy and a take-over of former Egyptian possessions, effectively turning the Red Sea into a 'British Lake' and fortifying key harbours along the eastern and southern edges of the Arabian Peninsula – the Trucial Sheikhdoms and Aden – to further secure the route to and from India (Onley, 2009).

This process had begun before the formal 1882 British occupation of Egypt but was deepened when French encroachment on the sources of the Nile persuaded Prime Minister Salisbury to restore imperial mastery over Sudan. Cairo had lost its grip over its Ethiopian and Sudanese possessions but rumours of possible French dam-building in the Upper Nile led Salisbury to dispatch Lord Kitchener with an invading force to Sudan, establishing the Anglo-Egyptian Condominium. With the fortification of Aden, the protectorate over British Somaliland (1886) and Uganda (1894), treaties with Italy (1891) and Ethiopia (1902) that recognized British control over the Nile and the conquest of Sudan (1898), the basin and the coastal supply points to and from the Subcontinent were safe. Britain's proconsuls tried to neutralize any threat to imperial shipping lanes, precluding autonomous Red Sea policies by its Arab or African clients; they also restructured political economies on the Peninsula and in the Basin to suit imperial needs (Darwin, 2012). Following ideas and practices pioneered under the Raj in India and Muhammad Ali (Gilmartin, 1994), British 'hydrocrats' identified big irrigation schemes and hydraulic infrastructure in Egypt and Sudan as essential to the material sustainability of the empire, helping to pay for occupation and supplying global markets with cotton and other cash crops.

London's hydro-hegemonic attempts lasted about three-quarters of a century (Tvedt, 2004). Imperial heroes such as Murdoch MacDonald dreamed of total 'Nile Control' through meticulous measurements, conservation and high productivity agriculture. MacDonald's 1920 master plan was updated by another British mandarin, Harold Edwin Hurst, who generated an all-encompassing plan to regulate the Nile flood from Ethiopia's Lake Tana and the equatorial lakes of Uganda and Congo to the river's Mediterranean mouth through a series of dams that could withstand extreme weather variation (Hurst *et al.*, 1946). As Northeast Africa and the Arabian Peninsula moved out of the British imperial orbit in the 1940s and 1950s, both the colonially crafted hydro-hegemony and dreams of a technocratic managerialism were bequeathed to post-independence elites in Egypt and Sudan. But while in Khartoum the new rulers were sectarian forces such as the Ansar and Khatmiyya who had been the main beneficiaries of colonialism and favoured a continuation of the (hydro) political-economic status-quo, in Cairo the Free Officers who came to power in their 1952 coup clamoured for a revolution to give Egypt not just political but also economic independence.

The fragmentation of British 'Nile Control' into different sovereign states – Sudan, Kenya, Uganda – with competing claims over the river risked pitting these newly independent countries against one another and leaving hopes of rapid development stillborn. Moreover, Gamal Abdel Nasser, the Free Officers' dashing leader, understood that passivity threatened to erode Egypt's advantageous position within the region: Cairo should act now that it still held the upper hand in the basin and was the only possible successor to British hydro-hegemony. Following his domestic land reforms, Nasser thus launched two inter-related projects to simultaneously affirm Egypt's regional leadership and to allow it to begin catching up with the West: objectives similar to Muhammad Ali's dreams 150 years earlier (Verhoeven, 2015b). His first priority was the Aswan High Dam which sought to centrally regulate the Nile flood for the first time in history, enabling predictable year-round irrigation and giving Egypt sovereignty over the waters. Complementing Aswan's infrastructural lock-in of Egyptian hegemony was the legal architecture of the updated NWA, which in 1959 volumetrically divided the available Nile flows between Egypt and Sudan, to the exclusion of other riparians.

Nowhere did this 'double lock' on Egyptian hydro-hegemony cause greater discontent than in Ethiopia. Emperor Haile Selassie officially maintained cordial relations with Egypt but felt that he was being presented with a long-term strait-jacket that demonstrated the hollowness of Nasser's professed Pan-Africanism: Egyptian national interests trumped other considerations. The Nile issue was moreover but one of multiple points of friction. The Free Officers sought radical autonomy from the West (cf. the 1956 Suez Crisis) and appealed to Arab nationalism across North Africa and the Middle East, rallying support among the monarchies of the Arabian Peninsula who backed the nationalization of the canal and heralded the Aswan Dam project as a victory for all Arabs. But while Nasser and the Gulf Arabs were keen to push the British and French out of the Middle East and Northeast Africa, Haile Selassie maintained good ties with Washington and London (Zewde, 2001). These alliances were partly a function of the fact that he had spent World War II in Bath and that British forces had liberated Ethiopia, as well enabled it to federate with Eritrea, which in practice meant an Ethiopian annexation of the former Italian colony.

In addition to maintaining a radically different relationship with Western powers than Nasser, the Emperor also feared Pan-Arabist rhetoric: at the aegis of an empire of dozens of nations and nationalities, Haile Selassie fretted over the depiction of Ethiopia as a 'Christian kingdom' subjugating its Muslim populations and denying self-determination to the Eritreans – a destabilizing discourse that could reinforce centrifugal tendencies. In the minds of Gulf Arab royals, the politics of the Nile Basin had an important ideological component: not only was the alliance with Egypt a bulwark against Zionist Israel, the pressure it simultaneously put on Addis was seen to help rectify a historical error. Wahhabi scholars and princes perceived the Ethiopian empire as a repressive obstacle to the natural penetration of Islam into the heart of Africa. Ethiopia's 'Biblical' alliance with Israel was portrayed as a Judeo-Christian challenge to the Muslim character of this key part of the world. As under Muhammad Ali and under British rule, any encroachment on the Nile risked destabilizing Egypt and thereby the dominant position of Islam at this geopolitical crossroads.

The geopolitics of the Nile and the perception of a 'Christian' Ethiopia helped Nasser claim the mantle of leadership of the Arab world. Seen from Addis, Egypt's bid for 'Arab mastery' over the basin was both a political and a civilizational challenge; a cold war developed between Ethiopia and the Arab camp, led by Cairo and Riyadh, in which each supported the other's rivals (Aalen, 2014: 627–628). Nasser and his Sudanese counterpart General Ibrahim Abboud sympathized with Somali claims to Ogaden in Eastern Ethiopia and backed the (mostly Muslim-led) Eritrean Liberation Front in the 1960s (Markakis, 1987: 102, 133), hoping that Eritrean secession would help turn the Red Sea from a British into an 'Arab Lake'. Haile Selassie retaliated by drawing closer to Israel, deepening his military alliance with the United States, which operated important intelligence facilities from the Kagnew base in Asmara and aiding the Southern Sudanese Anyanya rebels (Yihun, 2013: 36–40).

The stand-off between Ethiopia and Egypt, Sudan and the Gulf states lost some of its intensity as Haile Selassie brokered the 1972 Addis Ababa Agreement between Khartoum and the Anyanya rebels and the Arab camp disintegrated. Crucially, Cairo and Riyadh fell out: the Saudis had never liked Nasser's socialist bent and his republicanism veiled a regime change agenda that targeted 'reactionary' monarchies from Jordan to Yemen. King Saud (1953–1964) and Faisal (1964–1975) feared Nasser's United Arab Republic with Syria which they incessantly undermined; they also turned Yemen's civil war into Egypt's Vietnam. Nasser committed Egyptian forces to quash the Zaidi monarchy and give Cairo a key geopolitical outpost in the southern Arabian Peninsula from whence to influence both Gulf and Nile Basin politics. The Free Officers' gambit was clearly influenced by the British strategy of controlling the Red Sea; pro-Egyptian governments in Somalia, Sudan and Yemen pressured simultaneously Ethiopia and Saudi Arabia, but goaded a costly backlash as Gulf monarchies bankrolled the royalist Yemeni guerrilla. More than 10,000 Egyptians died before Cairo beat an ignominious retreat (Witty, 2001). Together with the crushing of the Egyptian army in the 1967 Six Day War against Israel, this spelled the end of Nasser's grand ambitions and the conflict between Arab 'progressives' and 'reactionaries' (Doran, 2006: 112–114). The Egyptian armed forces were demoralized and the economy stagnated: because Nasser's foreign image and his domestic hegemony were so interconnected, the twin blows Israel and Saudi Arabia had dished out decisively reshuffled regional politics.

The shifting sands of the 1970s: the rise of the Gulf and the gambles of Sadat and Nimeiri

As Anwar Sadat appeared at the helm of the Arab Republic of Egypt in 1970, he inherited a geopolitical situation in flux. On the one hand, Cairo was still the Nile's hydro-hegemon. Sudan remained pro-Egyptian, particularly under its Nasserist dictator Ja'afar Nimeiri, and Siyad Barre's Somalia helped to continue isolating its Nile rival Ethiopia. But on the other, Egypt was losing the leadership of the Arab world; the humiliations of Yemen and the 1967 defeat catalysed Cairo's decline, perhaps best expressed in Sadat's admission that because of its wars and rapid population growth, Egypt had become the poorest Arab state per capita. In the first years of his reign, the politically feeble Sadat failed to rein in spending (high patronage expenses were necessary to keep him in power) and the confrontation with Israel endured, further inflating the defence budget – resulting in an ever more dire economic outlook. The 1973 Yom Kippur war ended in a stalemate and the partial restoration of Egyptian prestige, but Sadat knew that debts were growing at more than 25 per cent annually: the origins of his U-turn in foreign policy and domestic economics (Dessouki, 1984). In order to restore Cairo's reputation and to avert a possible challenge from Ethiopia where a Marxist–Leninist military regime had taken power in 1974, Sadat launched his *al-infitah* policy of deregulation, liberalization and wooing foreign investors. Simultaneously, he restored diplomatic relations with Washington and, controversially, went to Israel in 1977 and signed the Camp David peace agreement with Menachem Begin in 1978.

This total reset intended to fortify Sadat's personal position at the expense of the army (Kandil, 2012: 113–163), to revive economic growth and for Egypt to catch its second breathe to secure its hegemony over the Nile and reassume Arab leadership in the 1980s.

In addition to Washington, the key partners in Sadat's stratagem were Gulf Arabs: the rising force in Middle Eastern and, increasingly, Nile Basin politics. The Gulf states started commercial oil exports in the late 1940s, with production soaring in the 1950s and 1960s and revenues reaching stratospheric levels owing to the 1973–1974 price hikes that resulted from the decision of the Organization of the Petroleum Exporting Countries (OPEC) to punish the West for its support for Israel. This incredible injection of wealth not only enabled Gulf countries to manage a spectacular socio-economic transformation without major political upheaval domestically, it also brought hundreds of thousands of Egyptian and Sudanese professionals and low-skilled workers to Saudi Arabia, Kuwait, the UAE and Qatar (Held and Ulrichsen, 2012). Both Egyptian and Sudanese expatriates and their home states effectively became dependent for their survival on Gulf rents. Consider Egypt first. Sadat's *al-infitah* was based on a simple economic formula: Egyptian cheap labour could be made more productive with Western technology and petrodollars. Support for Egypt was no longer a matter of mere diplomatic Arab declarations; between 1971 and 1977, Gulf financiers essentially covered the chronic balance of payment deficit Cairo kept on running. Yet while generating some growth, the gains of *al-infitah* were spread extremely unevenly and led to social polarization. Moreover, Sadat's political-economic pivot marked the beginning of a long-term dependency in Egypt on Gulf money that ended Cairo's project of Pan-Arabism and has limited Cairo's autonomy ever since (Ajami, 1978), including in the all-important policy arena of Nile Basin politics.

The rise of Saudi Arabia and Kuwait in particular also transformed Sudanese politics and Sudan's role in the region. For the first 15 years after independence, the traditional sectarian elites continued their business-as-usual management of the economy and were content with relatively modest regional involvement. However, as Ja'afar Nimeiri grabbed power in a coup in 1969, he launched the 'Breadbasket' strategy which promised that Sudan would feed the Middle East and Africa provided it could massively expand and upgrade its agriculture, with Western technology and Gulf money (Khalid, 1985: 62–66). The Arabian Peninsula had grown increasingly concerned about a widening 'food gap' – a function of skyrocketing imports and the failure of domestic cereal production to keep up with population and consumption growth – and Sudan with its potentially 'two hundred million acres of unused arable land' appeared an obvious partner: a political ally, a cultural cousin and an economic opportunity. Investment schemes were designed totalling billions of dollars over a 25-year time horizon: the sky seemed the limit (Woertz, 2013b: 161–182). For Nimeiri, the massive expansion of mechanised agriculture offered the possibility of both a cash bonanza and a political-economic makeover of Sudan. The Breadbasket was imagined to lead to the rise of a new economic class, dependent on the president's policies and patronage opportunities, which would consolidate Nimeiri's long-term hold over Sudanese politics and break the strong ties in rural Sudan between the sectarian forces and peasants (Verhoeven, 2011a).

However, the growing prominence of Saudi Arabia, especially in the 1970s and 1980s, also had more subversive manifestations from the standpoint of regime stability. Signing peace with Israel infuriated many across the Gulf – Cairo had left Riyadh almost totally in the dark – not least when intelligence reports indicated that Sadat was willing to grease the détente with Tel Aviv by selling possibly as much as 800 million cubic metres of Nile water a year to the Zionist enemy (Bulloch and Darwish, 1993: 79–84). Sadat's final years in office ended in acrimonious relations with Riyadh which scaled back its financial assistance, deepening the quagmire Egypt was sinking into. In this context of worsening malaise, Egyptian and Sudanese migrants to the Gulf comprised a lifeline for their home societies (Bernal, 1991). This economic dependence was complemented by accelerating social and political change induced by the intensification of exchanges between Nile Basin countries and the Arabian Peninsula. A powerful challenge to the Egyptian authorities arose owing to mounting ideological inspiration and financial support for Islamist dissidents of various sorts from official and private Gulf Arab sources – a trend all the more explosive because Cairo's peace deal with Israel and alliance with Washington were fanning the flames encroaching on Sadat's throne. The Egyptian president told King Khalid that he had no lessons to take about dealing with enemies when Saudi Arabia itself appeared besieged (cf. the millenarian extremists who attacked *Masjid Al-Haram* in Mecca in 1979). Religious fundamentalism increasingly threatened incumbents on both sides of the Red Sea, as Islamists penetrated elite and working-class circles alike; the man who in 1981 assassinated Sadat, Khaled Al-Islambouli, by no means represented a marginal faction of society but embodied a wider shift in Egyptian society (Heikal, 1983).

If in Egypt groups such as the Muslim Brotherhood, Al-Gama'a Al-Islamiyyah and Takfir wa Al-Hijra challenged the state from below, in Sudan the surging economic and ideological clout from the Gulf above all made itself felt at the level of the elites: so much so that it would help catapult the Al-Harakat Al-Islamiyyah (HI) of Hassan Al-Turabi to governmental power. Sudan's HI was disproportionately comprised of intellectuals and professionals who attributed the country's failure to realize the dreams of independence to the corrupt, anti-democratic and morally degenerate sectarian forces that had dominated Sudanese society and politics (El-Affendi, 1991). They admired Nimeiri's assault on these traditionalists and despite their early rivalry with the left-leaning president, Turabi and his lieutenants joined several of his cabinets in the late 1970s and became prime beneficiaries of the Breadbasket bonanza. In Kuwait, the UAE and Saudi Arabia, high-flying Sudanese émigrés with Islamist sympathies occupied important positions in banking, education, construction and agro-industry. The HI's privileged contacts in the Gulf meant that it could offer both established businessmen and young graduates unprecedented investment and credit opportunities. A new upper and lower middle class emerged in Sudan, interacting symbiotically with Turabi's HI: as the movement grew more popular and wealthier due to support from inside and outside Sudan, it could offer its prospective supporters a more attractive vision of material progress and religious piety; in turn, the sense that the HI was tomorrow's irresistible political force drew money and supporters displaying overt religiosity to it. Even when the decline in oil prices of the 1980s limited Saudi ambitions and Nimeiri fell amidst popular protests and the collapse of the Breadbasket in 1985, the Islamists – who had risen to prominence on the back of the Nimeiri–Gulf alliance – appeared to escape untarnished.

Unexpected defiance: the Al-Ingaz challenge to Saudi–Egyptian leadership

The 1970s was the decade during which Gulf states – especially Saudi Arabia – soared in international importance and acquired a regional role that was new: their underwriting of gambits for the consolidation of power by Egyptian and Sudanese incumbents and for confronting possible challenges by Arab and African communists (e.g. from Ethiopia) made them kingmakers of sorts in key Nile Basin states and with regards to the regional balance of power. But in the 1980s and 1990s, the Gulf Arabs would be confronted with the other side of the coin of influence: growing exposure also meant growing disappointments (Ehteshami, 2012: 262–263). The drop in oil prices, the Islamic Revolution in Shia Iran (and its export to Lebanon) and Israeli military supremacy reminded the Gulf states of their fragility in shaping the world around them. Sudan revealed itself as a particular, unexpected headache after having shown such promise in the 1970s.

Traditionally, Saudi relations with the Nile Basin have revolved around two basic geopolitical axes: the alliance-cum-rivalry with Egypt for leadership in the Arab world and distrust of 'Christian' Ethiopia (Verhoeven and Woertz, 2016). The latter appeared as especially dangerous between 1974 and 1991, because its Marxist–Leninist military regime was Soviet backed and in turn supported communist-inspired, secularist rebels from Sudan's marginalized peripheries who did not consider the country 'Arab', but fundamentally African.

This alliance threatened both the Sudanese government and Egypt's interests, as Cairo feared regime change in Khartoum or that a secession of Southern Sudan would undermine its dominance in the Nile Basin. Seen from Riyadh, Sudan had been regarded as the most predictable of the three dominant basin states and the most reliable in its dependency on Saudi Arabia and Kuwait. This perception changed drastically in the 1980s and 1990s because of two main factors. The first was the utter failure of the Breadbasket: not just from the perspective of thousands of Sudanese smallholders who were driven off their lands to enable mechanized agriculture in South Kordofan and the Blue Nile, but also for Arab investors who were disgusted by meagre returns and erratic behaviour by the Sudanese authorities. Many felt that they had been taken for a ride by the Nimeiri regime and its cronies: Sudan was still decades away from being the Arabian Peninsula's Breadbasket and, as oil prices dropped and money became tighter in Gulf states, such risky gambles could no longer be countenanced.

The second and most important reason for the changing assessment of Nile Basin politics by Gulf Arab policy-makers was that the Islamists, whose stellar rise in Sudanese politics they had aided, turned out to be a rather different kettle of fish than imagined. In 1989, Hassan Al-Turabi masterminded a *coup d'état*, which initially appeared to be a traditional military takeover without any ideological or international-revisionist agenda. However, when Saddam Hussein invaded Kuwait in August 1990, the new order in Sudan – which called itself 'Al-Ingaz' (the Salvation) – revealed its true colours: Turabi refused to join the US and Saudi led alliance to stop Saddam and called on Arab solidarity to settle the Iraqi–Kuwaiti dispute. The Saudis now grasped that behind the military façade of the 1989 putsch had stood the formidably well-organized HI which aimed for nothing less than a Sunni Islamic revolution in Sudan and condemned the lethargy and depravity of the status-quo in Cairo and Riyadh (Karawan, 1992). When military-Islamist Khartoum subsequently also invited Saudi dissident Osama Bin Laden to Sudan, alongside Egyptian radicals who sought the overthrow of Sadat's successor Hosni Mubarak and nearly succeeded in assassinating the latter in June 1995 with the help of Al-Ingaz intelligence services, a red line was crossed. Kuwait and Saudi Arabia organized a nigh-total boycott of Gulf investment and official development assistance (ODA); almost overnight the Sudanese economy was drained of much needed cash. Just when Ethiopia seemed to have become much less of a worry (the secession of Eritrea in 1993 deprived it of access to the Red Sea and the Marxist-Leninist regime had fallen in 1991), Sudan emerged as an indomitable challenger. Riyadh, Abu Dhabi, Doha and Kuwait City decided to balance against Al-Ingaz and drew closer to Mubarak's Cairo again, forgiving US$6 billion in outstanding loans for its participation in the Gulf War (Greenhouse, 1991).

From the perspective of Nile Basin politics, the Al-Ingaz regime's export of the revolution had the effect of assembling an *ad hoc* coalition of states that had historically distrusted each other and thwarted each other's plans: after the failed 1995 assassination attempt against Mubarak, Egypt and Ethiopia coalesced, partly with American help, in an anti-Ingaz front that also comprised Uganda and Eritrea to jointly work for regime change in Khartoum. The isolation of Turabi and Saudi financial engagement with Cairo amounted to a regional configuration that ostensibly strengthened Egypt's hydro-hegemony. The Mubarak state remained relatively weak and besieged by jihadist terror, but with Sudan deprived of hydro-agricultural investment and Ethiopia left without access to the Red Sea and still finding its feet after two decades of debilitating civil war, no challenges to Egypt's consumption of the overwhelming share of Nile waters seemed on the horizon. Mubarak's regime pushed on with new plans to green the desert and relaunch food exports: American and Gulf investors, as well as regime cronies, received generous concessions. The combination of liberalizing land tenure regulations after 1992 (Saad, 1999) and the controversial Toshka scheme that promised to reclaim millions of acres (Bush, 2007; Warner, 2013) simultaneously intended to buttress the patronage networks of the regime as well as to consolidate the dominant hydro-agricultural patterns of production and consumption – and political power – in the basin. An army largely depoliticized since Sadat's peace with Israel has increasingly been involved in rent-seeking in the agricultural and hydro-infrastructural sector, benefiting handsomely from land reclamation fantasies *à la* Toshka in which army companies play a highly lucrative role (Frisch, 2001). For almost a decade, high-profile Gulf engagement with the Nile Basin was largely restricted to the partnership of Riyadh and other capitals with Cairo.

Just how pivotal the Gulf states have become to the balance of power in the Nile Basin and the domestic fortunes of regional governments was underlined when these ten years of political-economic distancing between the Arabian Peninsula and Sudan were suddenly reversed. In December 1999, Al-Ingaz godfather Hassan Al-Turabi lost an internal power struggle against disgruntled Islamists and traditionalist army generals who wanted to abandon Sudan's tacit democratization efforts and its revolutionary foreign policy (Verhoeven, 2013a). President Omar Al-Bashir and Vice-President Ali Osman Taha immediately offered Egypt, Saudi Arabia, Kuwait and the Emirates rapprochement: an apology for the Gulf War error of judgement and expelling all dissident radicals in exchange for a normalization of ties and a return of Gulf investment and aid. The price to pay for defanging Khartoum was the renewed, massive influx of credit which enabled the revamped Al-Ingaz regime to launch its 'hydro-agricultural mission', which consisted of a multi-billion dollar dam programme – with the Merowe Dam as the jewel in the crown – and the Agricultural Revival Programme. The rationale was leveraging foreign investment and Sudan's first petrodollars to improve water control and transform agricultural productivity with a view to the electrification of urban and key rural areas and to export unprecedented quantities of cash crops. This updated version of Nimeiri's Breadbasket relies on a strong faith in capital-intensive agriculture and the recycling of rents in hydro-agricultural infrastructure – in the belief that this will both generate economic growth and build a new political class of pro-Ingaz loyalists who owe their material progress to the regime's benign governance.

The key involvement in Al-Ingaz' hydro-agricultural mission of Gulf actors through a mixture of personal connections, ideology and familiarity is perhaps best illustrated through the role played by Abdellatif Youssef Al-Hamad, a legendary figure in Gulf finance. Al-Hamad was educated in America as part of Kuwait's post-independence elite that tried to build the state, drawing on the country's astounding oil reserves. The devout Muslim and Pan-Arabist Al-Hamad became the founding director-general of the Kuwait Development Fund and built up a dazzling network of contacts within Western *haute finance* as well as among bankers, officials and entrepreneurs from Africa and the Middle East. His role as bank director in London put him in touch with Abdelrahim Hamdi, who would become Turabi's financial guru in the 1990s, but worked in Britain at the time and was active in the UAE. Al-Hamad developed a soft spot for Sudan and surrounded himself with several Sudanese assistants; his friendship with Hamdi and other Sudanese Islamists led him to instrumentalize his access to the highest levels – he has long been close to Kuwait's ruling Al-Sabah dynasty – to fund ambitious projects in Sudan. It was under Al-Hamad's 1972–1981 reign at the Arab Fund for Economic and Social Development (AFESD) that the AFESD became one of the chief architects of Sudan's Breadbasket gamble: its technocratic help, its articulation of Gulf Arab priorities and its role in channelling hundreds of millions to Sudan proved vital. As the Breadbasket turned sour, Al-Hamad became Kuwait's minister of finance and remained a confidant of the emir; he returned as AFESD director-general in 1985, a function he still retained when after the turbulent 1990s, Omar Al-Bashir reached out to Egypt and the Gulf, suggesting rapprochement. Bashir reappointed Abdelrahim Hamdi in 2001 for a second spell as finance minister (he had earlier served under Turabi) and the latter leveraged his excellent relations in the Gulf, particularly with Al-Hamad, to get the hydro-agricultural mission funded.

Hamdi and Sabir Mohamed Hassan, the governor of Sudan's Central Bank and another old friend of Al-Hamad, put the case to Gulf financiers that in order to lock in the de-radicalization of Sudan, to keep the 'extremist' Turabi out and to finally develop the country's indisputable agricultural and water resources potential, investment, aid and credit flows would be essential. The revamped Al-Ingaz signalled that it was open for business; the food and water security needs of the Arabian Peninsula had only sharpened throughout the 1990s and, as Hamdi and Sabir knew, in the minds of the Gulf Arabs, Sudan remained the inevitable, obvious partner to help remedy the worsening crises. The massive return of Gulf money should usher in win–win partnerships in which military-Islamist Sudan scaled unprecedented heights of prosperity and the Gulf's food deficit was mitigated: under the impulse of Al-Hamad, Hamdi and Sabir, bilateral and multilateral Gulf funds such as the AFESD committed to (and transferred) several billion dollars in support for dams at Merowe, Roseires and the Upper Atbara. Saudi, Qatari, Emirati and Kuwaiti investors devised ambitious plans; like in the 1970s, the sky seemed the limit.

The upshot of such high-profile investment by Gulf elites was that Al-Ingaz' hydro-agricultural mission materially changed the equilibrium in the Nile Basin. Because Sudan had also cultivated an alliance with China – which built Sudan's oil industry and assumed a leading role in the dam programme – and Egypt had been compelled to green light the development of Sudanese water and agricultural resources, the first major hydro-infrastructure initiatives on the Nile in four decades were undertaken. This watershed inevitably eroded Egypt's previously unassailable position – even if the Merowe Dam helps deal with the sedimentation problem at Aswan – not least as it also opened the door for an Ethiopian dam programme (Verhoeven, 2013b). For decades, civil wars, resource scarcity, isolation from international financing mechanisms and Egypt's long-arm dominance had inhibited any challenges to the hydropolitical status-quo from Sudan and Ethiopia. The arrival of China (Large and Patey,

2011) and the return of the Gulf states in the basin offered unprecedented financial and technical opportunities to states upstream from Egypt, as well as, in the outside world, a heightened awareness of intra-basin resources and how they could serve the interests of extra-basin suitors.

Conclusion: Gulf actors in the geopolitics of the Nile Basin today

This chapter has provided an overview of how Gulf players have been co-constitutive of the turbulent geopolitics of the Nile Basin for decades and have affected both the development of agricultural and water resources as well as struggles over domestic and international hegemony. The rise of the Gulf states was a function of both their soaring incomes from hydrocarbons in the 1960s and 1970s, but also of the institutional and infrastructural failures of the traditional hegemons in the Middle East – Egypt, Syria, Iraq and Iran – whose relative decline left a vacuum in the region and in adjacent geopolitical environments that has been filled by Saudi Arabia and, more recently, the UAE and Qatar (Kamrava, 2013: 2–3). Their financial prowess in the context of crises of state legitimacy and resource scarcity in the Nile Basin has equipped them with disproportionate clout to impact upon regional stability; moreover, in Sudan and Egypt, financial assistance from the Gulf has become a necessary condition to restructure political economies and redraw the hydro-agricultural landscape. Both Khartoum and Cairo have arguably been reduced to de facto clients of the Gulf.

This has been a particularly bitter pill to swallow for Egypt: Hosni Mubarak was propped up until his fall by Gulf states with whom he retained excellent relations throughout the 1990s and 2000s. To ordinary Egyptians, his regime seemed beholden to Washington, Tel Aviv and Riyadh, sinking into a foreign policy quagmire; his economic reforms mostly benefited the top sliver of society but barely had any positive effects on consumption patterns of the average household, as World Bank research shows (Verme, 2014). Soaring public debt, deepening inequalities and double-digit unemployment found their political corollary in a neoliberal police state where American and Gulf financial assistance kept the ship afloat, but rudderless (Mitchell, 2002: 270–303). An enfeebled Egypt largely disengaged from Africa and revealed itself as but a pale reflection of the formerly formidable hydro-hegemon. Trade relations with other African states are barely worth mentioning (none of them makes it into the top 15 of Egypt's export markets) and Cairo has been a bystander to sequential civil wars in Sudan and to the 2005 Comprehensive Peace Agreement that ultimately allowed South Sudan to secede – a turning point that Cairo tried to prevent for decades out of fear that it would erode its hydro-hegemony but was now powerless to thwart. The Egyptian government has been similarly stuck in a largely reactive mode to the construction of the Grand Ethiopian Renaissance Dam (GERD), Africa's biggest infrastructure project – a sure sign of how the basin's power relations have been overhauled. The 2011 Revolution offered an opportunity to reset Egypt's political economy and foreign relations but, following the overthrow of Muslim Brotherhood rule in 2013 and the authoritarian restoration, the regime of Abdel Fattah el-Sisi has survived on almost US$30 billion in support from Saudi Arabia and the UAE. The country's political and economic fortunes – and hence its role in Nile Basin politics – are more than ever dependent on Riyadh and Abu Dhabi.

This almost 180-degree reversal of what the power relationship between Egypt and the Gulf states was 60 years ago exemplifies the surging prominence of Gulf Arab actors in Northeast Africa and the Horn. But this should not be taken to imply that their potentially hugely consequential influence is also effectively and coherently leveraged: quite the contrary. Patterns of behaviour in the past decade underline two key features of the Gulf Arab–African relationship: deep divisions among the Gulf states and recurrent mutual disappointment between Gulf Arab patrons/financiers and African clients/partners.

The first observable trend is that of fragmentation: while Saudi Arabia's leadership was virtually uncontested until ten years ago, the north-eastern corner of Africa has since become a chess board on which Gulf regional rivalries are played out. Qatar and the UAE no longer simply follow Riyadh's 'Riyal Politik', but have their own interests, instruments and stratagems that sometimes overlap with those of the Al-Sauds but regularly also differ. This complex triangular relationship is further complicated by the intense rivalry between Saudi Arabia and Iran (Mason, 2015). The struggle for influence is fought on multiple battlefields: Tehran, Abu Dhabi, Riyadh and Doha pursue a mix of ideological, economic and security objectives and deploy hydrocarbon dollars, military assistance, media pressure, proxy forces and diaspora communities to outmanoeuvre each other. Sudan, a strategic crossroads at the heart of the Nile Basin, bordering Libya, Egypt, the Red Sea and Ethiopia, is an inevitable flashpoint.

Contrary to Egypt, which seems fixed in its geopolitical orientation and links to one Gulf patron or another (even during a year of Muslim Brotherhood rule, the Mohammad Mursi government was bankrolled by Qatar), what remains of the once revolutionary, military-Islamist Al-Ingaz regime currently pursues not a strategy of band-wagoning, but rather one of hedging in which it plays all of the above parties at once but allows none to completely dominate its patronage of Sudan. Thus, Khartoum maintains simultaneous economic, political and military relationships with Iran, Qatar, Saudi Arabia and the UAE and squeezes each of its patrons for maximum concessions and advantages. Iranian expertise has helped developed the Yarmouk military-industrial complex but Khartoum has also backed the Saudi-led military campaign in 2015 and dispatched troops against the Houthis in Yemen who receive support from Teheran's Revolutionary Guards. Al-Ingaz has weathered the storm of the loss of South Sudan and three-quarters of oil reserves via Qatar's policy of central bank to central bank support, but simultaneously courts Emirati and Kuwaiti investment in agriculture and real estate alongside Qatari projects. This careful playing off of various suitors-cum-patrons against each other – or 'managed multi-dependence' as one scholar coined it (Nonneman, 2004) – has weakened the political clout wielded by Gulf states and proven an effective survival mechanism for a beleaguered Omar Al-Bashir.

Al-Ingaz' skilful balancing underlines the second dominant characteristic of the Gulf Arabs' presence in today's Nile Basin: frustration. There is a barely hidden sense that for all the largesse that is bestowed upon Nile Basin clients, the results remain dispiriting, politically and economically. Gulf involvement in Sudan's Agricultural Revival was intended to help ease the immense water and food security pressures that the region faces as since the 1970s 'the Middle East has effectively ran out of water resources to become self-sufficient through domestic food production' (Allan, 2002: 9). However, the results have been dismal – like those of the Breadbasket 30 years earlier – with productivity languishing and dysfunctional partnerships with Sudanese counterparts grinding to a halt because of the same fundamental problems that have long bedevilled agricultural development in Sudan and elsewhere in Africa (Verhoeven, 2012). Thus, for all the non-governmental organization charges of wealthy Gulf Arabs importing thousands of tons of produce while displaced Africans languish in hunger, the bottom-line is that of a 'landgrab that wasn't', with very little food actually being exported and ambitious plans for Gulf investment more often than not coming to naught (Woertz, 2011, 2013a). This failure at a time of exacerbating water, energy and food security challenges strengthens calls for Arab investors to rethink their strategies and partnerships, for their own sake and those of the developing world's most marginalized populations (Keulertz and Woertz, 2015).

This balance sheet of economic disappointments is mirrored by political apprehensions. Gulf funding may have stopped Egypt and Sudan from imploding under revolutionary and insurgent pressures and thwarted the consolidation of power by forces antagonistic to Saudi Arabia in particular, but all the petrodollars spent have done little to strengthen the Egyptian and Sudanese state. In many ways, the transfers further contributed to their fragility and dysfunctionality. They certainly have not reduced illegal migration flows from the Nile Basin countries to the Gulf – an acute political headache – or enabled a better regulatory and institutional environment for Gulf investment, nor have they reversed the important influence gained by Iran (in Sudan) and Israel (in South Sudan, Eritrea and Ethiopia). Ethiopia, an old foe of the Gulf Arabs with persisting sketchy relations with the Peninsula, is powering ahead with the construction of the GERD, despite much Egyptian and Gulf Arab protestation. Riyadh, Abu Dhabi and Doha have little to no meaningful relationship with the Nile Basin's coming hegemon (Verhoeven, 2015a) and seem puzzled about how to rethink and redraw their strategic priorities and alliances in Africa's north-eastern corner. With Syria, Iraq and Yemen ensnarled in sectarian flames and structural economic and ecological challenges becoming ever more acute, falling back on the Gulf Cooperation Council (GCC) and default interactions with Egypt and Sudan appear inadequate. However, it remains to be seen whether the Gulf states can assume a qualitatively different role in the political economy of the Nile Basin: ideally, they would leverage their financial clout and substantial interests in the region to promote forms of symbiotic regional integration between basin states themselves and between the basin and the GCC. In light of the history dissected in this chapter, proving themselves willing and able to do so would certainly positively surprise many, including this author.

References

Aalen, L. (2014). Ethiopian state support to insurgency in Southern Sudan from 1962 to 1983: Local, regional and global connections. *Journal of Eastern African Studies*, 8: 626–641.

Ajami, F. (1978). The end of pan-Arabism. *Foreign Affairs*, 57, 2: 355–373.

Allan, J. A. (2002). *The Middle East Water Question*. London: I.B. Tauris.

Bernal, V. (1991). *Cultivating Workers: Peasants and Capitalism in a Sudanese Village*. New York: Columbia University Press.

Bowen, W. H. (2014). *The History of Saudi Arabia*. Second edition. Santa Barbara, CA: ABC-CLIO.

Bulloch, J. and Darwish, A. (1993). *Water Wars: Coming Conflicts in the Middle East*. London: Victor Gollancz.

Bush, R. (2007). Politics, power and poverty: Twenty years of agricultural reform and market liberalisation in Egypt. *Third World Quarterly*, 28: 1599–1615.

Cascão, A. E. (2009). Changing power relations in the Nile River Basin: Unilateralism vs. cooperation. *Water Alternatives*, 2: 245–268.

Darwin, J. (2012). *Unfinished Empire: The Global Expansion of Britain*. London: Allen Lane.

Dessouki, A. E. H. (1984). The primacy of economics: The foreign policy of Egypt. In: Korany, B. and Dessouki, A. E. H. (eds). *The Foreign Policies of Arab States*. Boulder, CO: Westview Press, 119–146.

Doran, M. (2006). Egypt: Pan-Arabism in historical context. In: Brown, L. C. (ed). *Diplomacy in the Middle East*. London: I.B. Tauris, 97–120.

Ehteshami, A. (2012). Security and strategic trends in the Middle East. In: Held, D. and Ulrichsen, K. C. (eds). *The Transformation of the Gulf: Politics, Economics and the Global Order*. Abingdon: Routledge, 261–277.

El-Affendi, A. (1991). *Turabi's Revolution: Islam and Power in Sudan*, London: Grey Seal.

Fahmy, K. (2002). *All the Pasha's Men: Mohamed Ali and the Making of Modern Egypt*. Cairo: American University in Cairo Press.

Frisch, F. (2001). Guns and butter in the Egyptian army. *Middle East Review of International Affairs*, 5, 2: 1–12.

Gilmartin, D. (1994). Scientific empire and imperial science: Colonialism and irrigation technology in the Indus basin. *Journal of Asian Studies*, 53: 1127–1149.

Greenhouse, S. (1991). Half of Egypt's $20.2 billion debt being forgiven by U.S. and Allies. *New York Times*, 27 May.

Halliday, F. (2005). *The Middle East in International Relations: Power, Politics and Ideology*. Cambridge: Cambridge University Press.

Heikal, M. H. (1983). *Autumn of Fury: The Assassination of Sadat*. London: Andre Deutsch.

Held, D. and Ulrichsen, K. C. (2012). Introduction. In: Held, D. and Ulrichsen, K. C. (eds). *The Transformation of the Gulf: Politics, Economics and the Global Order*. Abingdon: Routledge, 1–25.

Howell, P. P. and Allan, J. A. (1994). *The Nile: Sharing a Scarce Resource. A Historical and Technical Review of Water Management and of Economical and Legal Issues*. Cambridge: Cambridge University Press.

Hurst, H. E., Black, R. P. and Simaika, Y. M. (1946). *The Nile Basin Vol. VII: The Future Conservation of the Nile*. Cairo: Government Press.

Kamrava, M. (2013). *Qatar: Small State, Big Politics*. London: Hurst.

Kandil, H. (2012). *Soldiers, Spies, and Statesmen*. London: Verso.

Karawan, I. (1992). Monarchs, Mullas and Marshals: Islamic Regimes? *Annals of the American Academy of Political and Social Science*, 524: 103–119.

Keulertz, M. and Woertz, E. (2015). Financial challenges of the nexus: Pathways for investment in water, energy and agriculture in the Arab world. *International Journal of Water Resources Development*, 31: 312–325.

Khalid, M. (1985). *Nimeiri and the Revolution of Dis–May*. London: KPI.

Large, D and Patey, L. (2011). *Sudan Looks East*. Woodbridge: James Currey.

Lawson, F. (1992). *The Social Origins of Egyptian Expansionism During the Muhammad 'Ali Period*. New York: Columbia University Press.

Markakis, J. (1987). *National and Class Conflict in the Horn of Africa*. Cambridge: Cambridge University Press.

Mason, R. (2015). *Foreign Policy in Iran and Saudi Arabia: Economics and Diplomacy in the Middle East*. London: I.B. Tauris.

Mekonnen, D. Z. (2010). The Nile basin cooperative framework agreement negotiations and the adoption of a 'Water Security' paradigm: Flight into obscurity or a logical cul–de–sac? *European Journal of International Law*, 21: 421–440.

Mitchell, T. (1988). *Colonising Egypt*. Berkeley: University of California Press.

Mitchell, T. (2002). *Rule of Experts: Egypt, Techno–Politics, Modernity*. Berkeley: University of California Press.

Nonneman, G. (2004). Analyzing the foreign policies of the Middle East and North Africa: A conceptual framework. In: Nonneman, G. (ed). *Analyzing Middle Eastern Foreign Policies*. London: Routledge, 6–18.

Onley, J. (2009). *Britain and the Gulf Shaikhdoms, 1820*–1971. Doha: Center for International and Regional Studies, Georgetown University, Occasional Paper.

Saad, R. (1999). State, landlord, parliament and peasant: The story of the 1992 tenancy law in Egypt. In: Bowman, A. and Rogan, E. (eds). *Agriculture in Egypt: From Pharaonic to Modern Times*. Oxford: Proceedings of the British Academy, 387–404.

Salman, S. M. (2011). The new state of South Sudan and the hydro-politics of the Nile Basin. *Water International*, 36: 154–166.

Salman, S. M. (2013). The Nile basin cooperative framework agreement: A peacefully unfolding African spring? *Water International*, 38: 17–29.

Swain, A. (2002). The Nile River Basin Initiative: Too many cooks, too little broth. *SAIS Review of International Affairs*, 22: 293–308.

Tvedt, T. (2004). *The River Nile in the Age of the British: Political Ecology and the Quest for Economic Power*. London: I.B. Tauris.

Ulrichsen, K. C. (2011). Repositioning the GCC states in the changing global order. *Journal of Arabian Studies*, 1: 231–247.
Verhoeven, H. (2011a). Climate change, conflict and development in Sudan: Global neo-Malthusian narratives and local power struggles. *Development and Change*, 42: 679–707.
Verhoeven, H. (2011b). 'Dams are development': China, the Al-Ingaz regime and the political economy of the Sudanese Nile. In: Large, D. and Patey, L. (eds). *Sudan Looks East*. London: James Currey, 120–138.
Verhoeven, H. (2012). A regional breadbasket at last or another mirage in the desert? In: Allan, T., Keulertz, M. and Sujamo, S. (ed). *Handbook of Land and Water Grabs in Africa: Foreign Direct Investment and Food and Water Security*. London: Routledge, 41–54.
Verhoeven, H. (2013a). The rise and fall of Sudan's Al-Ingaz revolution: The transition from militarised Islamism to economic salvation and the comprehensive peace agreement. *Civil Wars*, 15: 118–140.
Verhoeven, H. (2013b). The politics of African energy development: Ethiopia's hydro-agricultural state-building strategy and clashing paradigms of water security. *Philos Trans A Math Phys Eng Sci*, 371: 20120411.
Verhoeven, H. (2015a). Africa's next hegemon. Behind Ethiopia's power plays. *Foreign Affairs*, 12 April.
Verhoeven, H. (2015b). The nexus as a political commodity: Agricultural development, water policy and elite rivalry in Egypt. *International Journal of Water Resources Development*, 31: 360–374.
Verhoeven, H. (2015c). *Water, Civilisation and Power in Sudan: The Political Economy of Military-Islamist State Building*. Cambridge: Cambridge University Press.
Verhoeven, H. and Woertz, E. (2016) Saudi Arabia, the Horn of Africa and the Geopolitics of the Red Sea. In: Partick N. (ed.). *Saudi Foreign Policy: Conflict and Co-operation*. London: I.B. Tauris.
Verme, P. (2014). Facts and perceptions of inequality. In: Verme, P., Milanovic, B., Al-Shawarby, S., El Tawila, S., Gadallah, M. and El-Majeed, E. A. A. (eds). *Inside Inequality in the Arab Republic of Egypt*. Washington, DC: World Bank, 55–100.
Warner, J. (2013). The Toshka mirage in the Egyptian desert: River diversion as political diversion. *Environmental Science & Policy*, 30: 102–112.
Waterbury, J. (1997). Between unilateralism and comprehensive accords: Modest steps toward cooperation in international river basins. *International Journal of Water Resources Development*, 13: 279–289.
Waterbury, J. (2002). *The Nile Basin: National Determinants of Collective Action*. New Haven, CT: Yale University Press.
Wiebe, K. (2001). The Nile River: Potential for conflict and cooperation in the face of water degradation. *Natural Resources Journal*, 41: 731–754.
Witty, D. M. (2001). A regular army in counterinsurgency operations: Egypt in North Yemen, 1962–1967. *Journal of Military History*, 65, 2: 401–439.
Woertz, E. (2011). Arab food, water, and the big landgrab that wasn't. *Brown Journal of World Affairs*, 18: 119–132.
Woertz, E. (2013a). The governance of Gulf agro-investments. *Globalizations*, 10: 87–104.
Woertz, E. (2013b). *Oil for Food: The Global Food Crisis and the Middle East*. Oxford: Oxford University Press.

Wu, X. and Whittington, D. (2006). Incentive compatibility and conflict resolution in international river basins: A case study of the Nile Basin. *Water Resources Research*, 42, 2.

Yihun, B. B. (2013). Ethiopia's role in South Sudan's march to independence. *African Studies Quarterly*, 14: 35–54.

Zewde, B. (2001). *A History of Modern Ethiopia, 1855–1991*. Oxford: James Currey.

Zeitoun, M. and Warner, J. (2006). Hydro-hegemony: A framework for analysis of trans-boundary water conflicts. *Water Policy*, 8: 435–460.

5 Inward investment in Sudan

The case of Qatar

Martin Keulertz

Introduction

Inward investment in the Nile Basin is not a new phenomenon and especially not in Sudan. What was described as 'land grabbing' in the global media was in fact the second attempt of Arab economies to utilize the perceived Sudanese 'breadbasket'. While the first wave that occurred in the 1970s and 1980s quickly faded, the second wave began in the late 1990s when the *Agricultural Revival* (*Al-Nahda Al-Zira'ayah*) programme was launched to invite agricultural investment from Islamic countries around the world (Verhoeven, 2013). Most of the analysis focuses on how land investments – by either emerging Asian economies or private-sector investors from across the globe – impact upon land use, land rights, water use, human rights and livelihoods (Cotula, 2011; Deng, 2011; Woodhouse and Ganho, 2011; Cotula, 2012; Anseeuw *et al.*, 2013; Woertz, 2013a, 2013b). Very few studies have taken an approach that analyses the role of international food politics and contentious domestic politics over investment (Paarlberg, 2010; Carmody, 2012; Brautigam, 2013; Woertz, 2013a, 2013b).

This chapter will present a failed investment in Sudan by Qatar. Despite significant media attention and signed contracts, the agricultural investment in Abu Hamad has not been implemented until 2016 (Land Matrix, 2016).[1] By discussing the political economic drivers of the State of Qatar behind the initial decision to invest in Sudanese agricultural land, the chapter will highlight the deeply political nature of agricultural investment – and its potential failure. The State of Qatar is the third richest country by gross domestic product (GDP) per capita in the world due to its natural gas wealth (International Monetary Fund, 2015). However, Qatar has limited capacity to produce its own food, due to its arid geographical location. This leads to legitimate food security concerns in Qatar. The Gulf state heavily depends on 'virtual water imports'[2] to provide food security to its citizens. It has to 'import' virtual water from the world market making the Gulf economy vulnerable to food price spikes as in 2007/2008 and 2010/2011. Qatar also has complex domestic political economic structures defined by 'shadow state' actors and their interests. Framing these domestic political structures in the context of inward investment in land is the objective of this chapter. In short, the chapter will discuss the role of food politics in Qatar and its crucial relation to inward investment in Sudan.

Setting the scene: food and water security in Qatar

The defining moment of food politics in the twenty-first century occurred during the 2008/2009 and 2010/2011 food price spikes, which caught the world by surprise (Piesse and Thirtle, 2009). It was especially the pace of food price increases that caused great concern among food importers. In particular, staple food such as wheat, sugar, rice, soy and other animal feed grains was affected by the spikes. Staples are normally produced on large-scale, rain-fed or irrigated land with low margins. Staples tend to be cheap thanks to economies of scale and highly developed supply chain management. However, few regions in the world have the capacity to produce staple food on large land areas – most importantly Latin America, North America and Northern Asia. Failing harvests in Russia, low storage in food producing countries, policy changes by the European Union and US governments to promote bio-ethanol as an alternative to fossil fuels, cuts in research and development budgets for agriculture by Western governments and food commodity speculation by financial markets led to unforeseen price increases. Between 2006 and 2008, average world prices for rice rose by 217 per cent, wheat by 136 per cent, corn by 125 per cent and soybeans by 107 per cent (Piesse and Thirtle, 2009). Despite the severe spikes, food was nevertheless relatively cheap compared to the price spikes in the 1970s when food hit dramatically high levels (Piesse and Thirtle, 2009).

However, the government in Qatar was nevertheless alarmed as 'there was no food on the market' as the former head of the Qatar National Food Security Programme (QNSFP), Fahad Al-Attiyah,[3] highlighted. All of a sudden, Qatar decision-makers were reminded about their dismal food security predicament, caused by the country's limited agro-hydrological situation and the dependence on the world food markets.

Qatar is located in a geographical position with an arid climate, virtually no rainfall and hardly any surface water. In hydrological terms, the State of Qatar is an absolute water pauper, subjected to severe physical water scarcity. Qatar receives a negligible amount of rainfall – about 100mm per year – confined to the winter months. In addition, Qatar has access to 0.058km^3 annually of renewable water largely from groundwater aquifers (FAOSTAT, 2013). About 8,000 people are associated with the agricultural sector that cultivates 15,000 hectares of land. The most recent GDP of Qatar is US$189 billion, which translates into a GDP per capita of US$102,000 (CIA Factbook, 2013).

Qatar had a population of 25.000 people in 1950. Thanks to an oil and gas boom, which generated an estimated US$90 billion net revenues in 2010-11, and the inward migration of professionals and labour workers from abroad, the population grew to 1.7 million in 2010 and is expected to grow to 2.6 million by the year 2050 (CIA Factbook, 2013). It must be noted that such predictions entirely depend on future economic growth fuelled by fossil energy as the Gulf state is dependent on oil- and gas-generated revenues and non-Qatari high- and low-skilled labour to keep the economy growing. The construction sector in the economy heavily relies on low- and semi-skilled foreign labour that generates substantial demand of imported staple foods. At present, the 1.7 million population in Qatar uses an average of 1,200 litres of water per day for domestic, industrial and recreational use, which is the highest consumption rate in the world. For drinking water purposes, Qatar has an emergency storage of only 3–5 days.[4]

Seventy per cent of all water is consumed by irrigated farming. Data on food consumption in Qatar confirm its dependency on agricultural imports (Figure 5.1). Nevertheless, the vast majority of food (90 per cent) is imported. Translated to water, the virtual water 'imports' of Qatar in the period from 1996–2005 have been estimated to 1,057 million cubic metres (mcm)/year (Figure 5.2). In terms of per capita food imports, Qatar is one of the largest food import dependent economies in the world. Import costs have sharply increased in the first decade of the twenty-first century. The import costs almost doubled for selected staple commodities in the period from 2007–2012. Since 2012, global food prices have gradually decreased. However, the five-year food price volatility bonanza sent out a warning message to the Qatari leadership that food security may be a bottleneck for political economic development. In case of high global food prices, increasing domestic agricultural production is not an option to mitigate food import dependency due to severe water constraints and the absence of technologies to produce crops with very limited water. In general terms, water is the main limiting factor in Qatar for the future prosperity of the country.

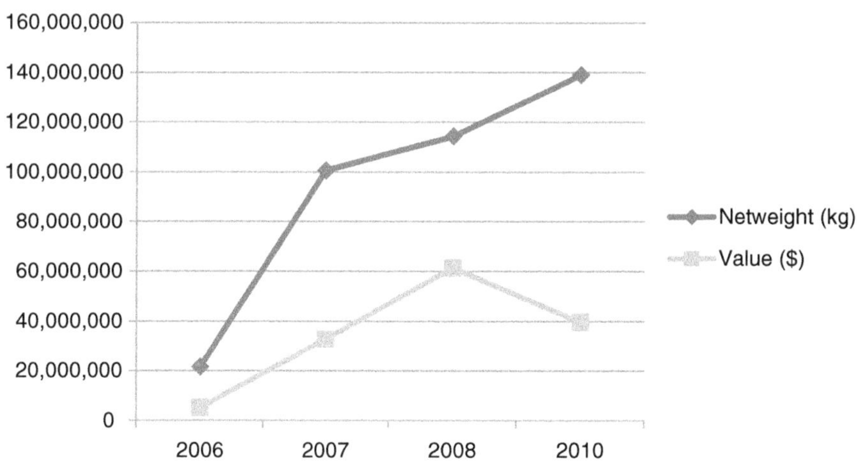

Figure 5.1 Qatar wheat imports from 2006–2010 in US$ value and kg quantity

Source: COMTRADE (2013).

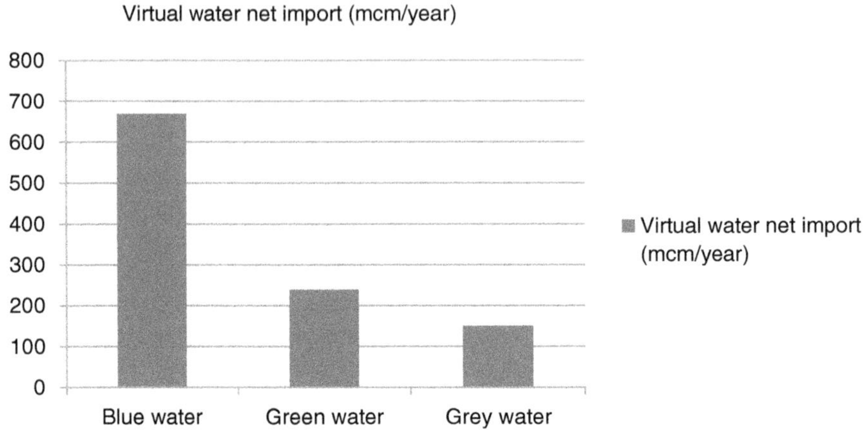

Figure 5.2 Qatar net virtual water imports in mcm per year

Source: Water Footprint Network (2012).

Sudan

Sudan is regarded as one of the few Arab countries that has not utilized its hydrologic endowments. However, when food prices were depressed, the invitation of the Sudanese government to Islamic governments offering land in the course of the *Agricultural Revival* (*Al-Nahda Al-Zira'ayah*) programme in the late 1990s was never taken up by Arab governments. For Sudan, agriculture is a potential source of foreign direct investment (FDI) to mitigate the impacts of a US-enacted Western trade embargo. In 1997, US President Clinton issued Executive Order 13067 to impose sanctions upon the Islamist government in Khartoum due to American intelligence reporting a proximity to international terrorism and violations of human rights by the Sudanese government (Embassy of the United States in Khartoum Sudan, 2015). Sudan had portrayed itself as the 'breadbasket' of the world since the 1970s to highlight the untapped agricultural potential along the banks of the River Nile (Waterbury, 2002). Although Sudan received 20 per cent of the Nile water in the colonial treaty brokered by the British in 1929, it never fully utilized its share. In Qatar, the Sudanese breadbasket was thus perceived as the potential remedy for food shortages in the whole MENA region.[5]

Qatar was allocated 100,000 hectares in Abu-Hamad by *Al-Nahda Al-Zira'ayah* in the late 1990s. However, Qatar only accepted the offer to invest in Sudan after the food price spikes in the post-2007 period. One reason for Qatar's reluctance to invest in Sudan in the early 2000s was the high costs of turning 'virgin land' into agricultural production. Agricultural investment experts reckon that an investor must be prepared to invest between US$1.3 billion and US$2.6 billion for 100,000 hectares in a country such as Sudan. In particular, the costs vary if gravity-fed or pumped irrigation schemes are favoured. Moreover, such investments require a corporate structure that will have to be hired internationally.[6] Deploying such large amounts of money for food production only makes sense if the drivers are understood. One important driver for Qatar's interest in land in the post-2008 period was the role of global food politics. Being in the position as a net-importer brings about more costs than just financial ones in the highly globalized world of food trade and food politics.

Global food trade and the potential political costs of dependency

Highly dependent food importers such as Qatar are both economically and politically dependent on food imports. However, global staple food production and trade is a highly strategic political resource. While large and small farmers produce food on their land around the world, the trade is in the hands of a few traders. Cereal traders such the ABCD in global food trade (Archers Daniels Midlands, Bunge, Cargill and Dreyfus) have 150 or more years of global trading expertise and they remain major players in the Western-led global staple grain supply chains (Murphy *et al.*, 2012). Subsequently, supermarket retailers and wholesalers rapidly grew in size and power as a consequence of deregulation and the arrival of computing and other logistics in the 1980s. By incorporating new regions (Latin America, Africa and Asia) into supply chains to provide more meat and fish, more fresh fruits and vegetables and more biofuels, Western consumers were able to experience an unprecedented nutritional journey. In addition, the industrialization of food production and the subsequent introduction of economies of scale and scope enabled traders and retailers to produce more food at lower prices (Teubal, 1993). The ABCD companies are experts in gaining access to natural resources such as water and fertile soils located in the so-called 'food bowls' of the world. The regions with food surpluses are geographically located in North and South America, Australia, the former Soviet Union and some countries in Eastern Europe and Asia. These are the regions from which Middle Eastern economies such as Qatar have imported their food since the twentieth century (Allan, 2002). The volumes of food imports have increased steadily over the past 100 years. It was estimated by Cargill that the Middle East as a region has alone imported 600 million tonnes of grain since the 1970s (*The Economist*, 2012).

This trade has allowed the Middle East to enjoy food and water security. The apparent security is meaningless, however, without an awareness of the 'nodes of power' within the agricultural and trading sectors of the world's 'food bowls'. The power relations in these 'food bowls' have dramatically shifted from public to private governance, especially since the early 1980s (Hendrickson *et al.*, 2008). At the same time, Asian traders are quickly catching up with Western market 'hegemons' using similar political economic strategies to those of their Western counterparts. New economic giants from East Asia, the NOW companies (Noble, Olam and Wilmar) reveal the vulnerability of the Middle East in the global food trade. All three companies have been strategically subsidised by Asian governments in China and Singapore to counter Western hegemony over food trade (Sojamo *et al.*, 2012; Keulertz and Woertz, 2015). These international drivers could even add more severe constraints on the political economic development of Qatar in the era of food globalization and population growth. In particular, Asia's growth and subsequent appetite for resources could push Arab government to the fringes in global trade.

Moreover, food trade is also decisively political. During the price spikes of 2007/2008, governments such as Qatar felt reminded of the 1970s. Food exports to the Middle East have been frequently used by the USA to conduct foreign policy with other means. What Woertz calls the 'food weapon' was very evident in the 1970s when the USA threatened to use the 'food weapon' during the heydays of the Organization of the Petroleum Exporting Countries (OPEC) oil embargo in 1973 as a response to Western support for Israel during the Yom Kippur war (Luttrell, 1981; Woertz, 2013b). The prevailing slogan underlining this threat was 'we freeze, they starve' together with the idea of establishing a 'grain cartel' (Luttrell, 1981).

Although food export boycotts were never used by the Americans against the Middle East, they did not fail to leave a trauma in the wider region, including Qatar. In January 1980, the USA imposed a food embargo on the Soviet Union by not selling 17 million tonnes after the dismal harvests following the 1980 drought in the Soviet Union (Paarlberg, 1980). Carter's foreign policy decision to place an embargo on the Soviets sought to penalize the Soviets for their invasion of Afghanistan. However, the Soviets acquired wheat from other world regions such as Europe and Latin America, and hence the effect was negligible on the Soviet Union.

In the Middle East, however, the use of the wheat embargo was perceived with severe anxiety. As one key informant stressed: 'Food insecurity has the potential to topple governments.'[7] Agricultural investors in Doha and Cairo stressed that food dependency on the West could be a dangerous strategy, hence the need for investment in Sudan.[8]

Thus, the past use of food as a foreign policy tool and Asia's rise on the world market explains why investment in Sudanese land was considered a viable policy choice by Qatari decision-makers to 'import' virtual water to counter the dismal domestic water predicament. However, there is another, largely untold layer in political decision-making, which is the domestic side in food politics.

The power of the shadow state and its influence on decision-making

Another important reason for deciding to invest in Sudanese land is owed to the complex history of Qatar that left its levy on the sociological composition of the modern economy. Qatar was a former British colony, yet with little significance due to their desert habitat. However, this changed after World

War I and more significantly due to the wider region's ascendancy as the prime provider of fossil energy to the global economy. The global political significance of the Middle East and thus Qatar as a country within the MENA region increased gradually thanks to access to oil and gas and the resulting financial assets. However, certain actors, referred to as the 'shadow state', benefitted both economically and politically from the increasing importance of the MENA region. A 'shadow state' is a neo-patrimonial form of rule that exists behind the official facade of laws and government institutions. Authority is based upon the decisions and interests of powerful individuals, not a set of written laws and procedures, even though these formal aspects of government may exist. The actors are tied to a ruler through a common regional background and through kinship of tribal affiliation (Tripp, 2007). These actors are also highly important in explaining the drive for land investments.

The Al-Thani ruling family have been in power since 1868 when Great Britain intervened politically to support the independence of Qatar from the Bahraini House of Khalifa. Bahrain controlled Qatar from 1783–1868. The British heavily influenced regional politics during the nineteenth century because the Persian Gulf was viewed by London to be strategically important in controlling the trade routes from India. However, due to the Ottoman Empire's geostrategic interests in the region, Qatar was forced to become part of the Ottoman Empire. Ottoman rule lasted from 1871–1916.

During World War I, the Al-Thani tribe took part in the Great Arab Revolt to expel the Ottomans from the Arabian Peninsula. Qatar once again found political shelter as a British protectorate until 1971 when the Gulf state gained independence (Crystal, 1990; Fromherz, 2012). Given limited oil wealth, Qatar had very little political significance for the British and the rest of the Western world. This situation changed, however, after 1971 when oil prices increased to unprecedented levels. During the 1980s, Qatar was affected by declining global oil prices as were the oil economies of the rest of the Arab peninsula, causing almost a decade of economic stagnation. However, since the 1990s, the energy sector has been transformed through natural gas development projects that have supplied gas to Europe and Japan. The roaring 1970s and the roaring 1990s enabled Qatar to transform its economy (Crystal, 1990; Fromherz, 2012). Energy-related developments induced an influx of over 1.5 million foreign professionals and workers. The small population of Qatar – the current indigenous population is only approximately 250,000 – required the ruling class to depend on foreign labour to maintain economic growth (Hanieh, 2011; Fromherz, 2012). Yet, the governance style of the state remained untouched.

The informal networks of power in Qatar

As in all Gulf States, the government of Qatar has kept tight control over the economy and no foreigner can set up a business without Qatari involvement. The size of the country and the long tradition of the House of Al-Thani have enabled the ruling class to control and manage the economy in a dirigiste, state-capitalist way where ownership is centralized and can be traced back to the House of Al-Thani (Hanieh, 2011; Fromherz, 2012). Despite the small size of Qatar, its social composition is profoundly tribal. The Al-Thani ruling family has been the largest tribe since its establishment in 1825. Its political power has always been flanked by a small group of merchant families who demand their share of the country's economic assets. Until 1971, the ruling family did not have much to share with its political competitors, and hence the developmental leap had to be shared carefully after the 1970s to allow the participation of all of Qatar's economic stakeholders (Fromherz, 2012). Foreigners are very subordinate in political terms in Qatar and have no right to interfere in the economy. The Qatari tribes have to be appeased by the Al-Thanis in order to gain a measure of political support for their rule.

The real nexus of power in Qatar is located in the inner circles of the royal family and its loyal merchant families[9] (Fromherz, 2012). Only men matter in political terms. Allegiance is not a political wild card; it has to be earned and deserved on an everyday basis. The former emir of Qatar, Hamad bin Khalifa Al Thani, who abdicated in 2013, governed his country like a CEO (Fromherz, 2012). He secured allegiance via financial handouts to Qatari citizens that involved 'subsidies, state jobs, land grants, and free university education, the latter part of an effort to reinvest gas wealth in human capital, not just in shiny buildings' (Kinninmont, 2013). The emir, the former crown prince and current emir and the former prime minister, who are all Al-Thanis, oversaw the $300 billion assets of the country and decided upon the money flows both within and outside Qatar.

Despite this rather hierarchical structure, the emir did not exercise his rule in an authoritarian way but rather pays close attention to *asabiyya* (tribal solidarity). Although the royal family is at the centre of the power web in Qatar, it has been possible for other merchant families to gain access to the royal patronage through marriage (Fromherz, 2012). For example, the second most important family in Qatar is the Al-Attiyah family because the mother of the emir is from this clan. Leading government members have a tradition to being sourced from this family, which allows the family a unique position in Qatari tribal relations.

The role of the Al-Attiyah family was significant in food security issues because high-profile jobs relating to food security and climate change were allocated to members of this family. Climate change politics have been high on the Qatari agenda because the small Gulf economy was host of the Conference of the Parties (COP) 18 meeting in December 2012. For Qatar, hosting the COP18 meeting was viewed with immense pride because it allowed the political leaders to gain access to the global political stage. Members of this family prepared and led the negotiations of the meeting. Parts of the Al-Attiyah family were part of what will be referred to as the environmental 'shadow state' comprising a young elite around the former crown prince and current emir that chose the environment as a topic of concern. Although agriculture is a negligible economic sector in Qatar, it was used as a political vehicle by the environmental shadow state to gain political influence over international issues such as climate change.

Two factions within the Qatari shadow state

Within Qatar's shadow state, two contending approaches to food security became apparent after 2009, which reflect the interests of the two conflicting factions of Qatar's 'shadow state'. First, there was a domestic objective in one Qatari plan that aims to increase food security gathered around the former crown prince and the Al-Attiyah family. Second, there was an international strategy to increase food security via overseas investment and trade gathered around the former Prime Minister Hamad bin Jassim Al-Thani. These two different political camps in Qatar had been assigned responsibility for the two strategies. The 'environmental shadow state' aimed at domestic solutions and the 'foreign policy shadow state' aimed at international solutions such as inward investment in Sudan.

Due to Qatar's dependence on food imports, the food price spikes in 2008/2009 prompted the crown prince to establish the QNSFP to find solutions to the country's growing food security concerns. The concern that drove the establishment was not 'the price of food in 2008/2009 but the limited availability of food on the global market that wasn't even sold to us (the Qataris) for an offered premium'.[10] As a result, the QNSFP was well endowed from the beginning of its operation, with a very large budget of approximately US$3 billion per annum to spend on food security policy measures and research. The cost of the expansion of domestic agriculture was estimated at $50 billion over ten years to reach a food-sufficiency rate of 60 per cent by 2023. The development of such a domestic supply chain would have required that very innovative technologies had to be deployed. They had to produce unprecedented volumes of food commodities and, at the same time, remedy the damage done to local aquifers by over-pumping for irrigation in the past. Finally, the production of food on the Qatari peninsula would also have provided landholders with opportunities to profit from investments in agricultural land, which is owned by other influential families in Qatar.

The QNSFP's original task was to identify strategies and policies to increase food security via domestic agricultural investment. The investment strategy focused on global food supply chains and, second, on agricultural land in Africa and other world regions. This last task was given to a company – Hassad Food – that was established as a subsidiary of the Qatar Investment Authority (QIA), which is a subsidiary of Qatar Holding where all the financial power of the emerging economy is located. Qatar Holding was chaired by the prime minister and foreign minister of Qatar, Hamad bin Jassim bin Jaber Al Thani, who was part of the 'foreign policy shadow state'.

Hassad is the agricultural trading and investment arm of the QIA. The objective of Hassad is to invest in global agricultural markets for the Qatari state, and differs substantially from that of the QNSFP. Its objective is to invest in companies all around the world that do business in the food supply chain. For this activity it has been provided with a budget of US$20 billion.[11] It is also the company that has been given the task of investing in global farmland, including farmland in Sudan. Like the QNSFP, it has been a newcomer to the field, having been established in 2008.

After 2011, the QNSFP and Hassad became caught up in internal 'shadow state' struggles. The contention has been over 'turf' issues and over who could exert power rather than on a rational review of natural resources use and development of, for example, water or of the uncertainties of high-risk ventures such as farmland development in economies with poor infrastructures. The most important aspect of this internal 'shadow state' struggle is Qatar's ambition to become a regional if not global political actor and to achieve its own food security in the process (QNSFP, 2013).

As a result of the carving up of political space, the allies of the crown prince, namely representatives of the second family in the state, were given a leading role during the highly prestigious COP18 meeting in Doha in December 2012. The conference series has assessed the progress of the United Nations Framework Convention on Climate Change on an annual basis. The deputy prime minister of Qatar and one of his nephews (from the Al-Attiyah family) were the key Qatari representatives. They chaired the meeting attended by international ministers of state in charge of environmental affairs. These members of the second family in the State of Qatar are part of the camp around the crown prince whose influence stems from the leading female figure in Qatar, the wife of Emir Hamid, who 'wanted her boy to be the main man in Qatar'.[12] In addition to the COP18 meeting in 2012, the crown prince's camp also organized a high-profile conference on Food Security in the Drylands in November 2012. In order to influence the range of choice on food security, the QNSFP organized this meeting to confirm the formation of the Global Dry Land Alliance (GDLA). The GDLA is an alliance of countries with poor water resources that need similar new technologies and farming systems to those needed by Qatar to improve and secure sustainable agricultural production.

The GDLA specifically targeted Middle Eastern countries such as Syria, Iraq and Egypt, where an increase in food production is seen as helpful for the Arab world, compared to global initiatives that compete with the experienced players in the global 'food regime'.[13] In convening the event, the organizers sought out well-known international political figures to support the agenda of the GDLA. Prominent international figures such as then US Secretary of State Hillary Clinton and UN General Secretary Ban Ki Moon were cultivated to give keynote presentations. However, Clinton and Moon eventually withdrew their initial acceptance of the invitation, which prompted the crown prince not to attend the meeting, despite earlier plans to address the audience.[14] Thus, it can be interpreted that the crown prince saw no value in getting personally involved in the conference as this attempt to send a message to the other camp around the prime minister failed.

At the same time, the foreign policy camp within Qatar's 'shadow state' viewed the conference with great caution. This camp around the prime minister became increasingly critical of the QNSFP's strategy, because it challenged his own portfolio, which handled international affairs and international investment. His camp was challenged by the GDLA initiative because it touched upon interests of the prime minister's camp. The prime minister wanted to control the billions of US dollars that have been made available by the emir for agriculture and food security after the food price spikes in 2008/2009. As a result, the QNSFP was closed down in 2013. The GDLA was given to a very senior Qatari foreign policy expert, who previously served as director general of the United Nations Economic and Social Commission for West Asia in Beirut. Bader Omar Al Dafer, however, never took the GDLA beyond concept stage. Meanwhile, the closure of the QNSFP and the downsizing of the GDLA did not affect Hassad Food. The prime minister's camp was not only interested in financial perks but in political influence because his camp was also the driving force behind Qatar's recent military intervention (e.g. in Libya) and their financial support for opposition groups in Syria and for the Muslim Brotherhood Government in Cairo and Sudan (Achcar, 2013; Khatib, 2013). Hassad Food was described as a tool for the prime minister's camp in its geopolitical ambitions to exert influence in North and East Africa.[15]

The geopolitical element: the role of water

This brings the chapter to the last point: the geopolitical sphere of Qatar's interest in Sudanese land. In areas such as Abu Hamad on the bend of the River Nile in central Sudan where a significant increase of irrigated vegetation cover was planned by the Qatari investors, the consequences for local downstream users would have been very significant. According to Hassad Food, a canal was constructed by the Qatari investors to divert 1bcm (billion cubic metres) of Nile water to the investment plot.[16] This suggests that for the Qatari project, this would have translated into a vegetation cover increase of over 50 per cent in the Abu Hamad area. Data from the WaterWorld model allowed for comparison between the water balance prior to the planned land use change and the potential impact of the project upon the water balance in the Abu Hamad area. It revealed that the project would have made use of up to 441.9mm/year of water resources in this area. This impact would have 'wiped out' the water balance completely in the project area. The Abu Hamad region has a water balance ranging from 0.9–1.74mm/year. The water 'gap' to irrigate the intended crops would have to be sourced from surface water such as the Nile, hence the project would also have had additional wider basin impacts (Mulligan, 2013).

Low rainfall of 4mm in August and September would have made the use of 'blue water' inevitable in the Hassad Food project. WaterWorld data indicates that the project in Abu Hamad would have posed severe risks not only to the project area but also to downstream populations. Increased virtual water 'flows' in the food produced on the Qatari project in Sudan would thus have been a major intervention in the lower Nile Basin.

Despite the assumptions of the Qatari investors, water was a limiting factor for agricultural production on the scale envisaged in Sudan. In a highly politicized basin such as the Nile, the affected downstream users are predominantly based in Sudan's neighbouring economy – Egypt. Downstream users in Egypt fully absorb all of its allocation and the additional water that Sudan is not yet utilizing from its nominal share for their agricultural sector to produce food for Egypt and its export markets. However, it is projected by an internal Egyptian ministerial study that demands in Egypt have further increased. Although the Egyptian government claims that Egypt only uses 56.8bcm, which would already be 1.3bcm more than allocated by the Nile Treaty of 1959, the updated projections of annual water use are dramatically higher. According to an internal study, Egypt will require 71bcm to meet its domestic, industrial and agricultural demands by 2017 (Ministry of Water and Irrigation of the Arab Republic of Egypt, 2011).

The alarming figures for Egyptian water demand would have been affected by the agricultural investment plans of Qatar in Sudan. The canal alone that is still constructed and ready to convey 1bcm per annum to irrigate alfalfa and wheat crops could exert further hydrological pressures on Egypt. The water of the Nile is already fully allocated in the Nile irrigated lowlands and delta, where the vast majority of Egypt's water is used for agricultural production.[17] The farmers in Egypt would have been seriously affected by the plans to expand Sudanese irrigated agriculture through the utilization of the perceived 'virtual basin'.

However, the last point of this chapter is to emphasize Qatar's role in global affairs and Middle Eastern political developments. Until 2013 when the Tamim was crowned as emir of Qatar and Prime Minister Hamad bin Jassim Al-Thani pulled the strings in Qatari foreign policy, Qatar's foreign policy was described as 'adventurous' and seeking to 'buy advantage' across the world to reinforce Qatar's geopolitical ambitions (Kerr, 2013). A cornerstone in the geopolitical ambitions of Qatar had been the QIA, of which Hassad Food is a subsidiary. The QIA used its economic power to invest in 'economic opportunities' and 'emerging trends' by investing in equities, private equity, real estate, alternative investments, special situations, strategic and direct investments, and commodities and natural resources (Khatib, 2013; SWF, 2013). The QIA portfolio includes property in London (The Shard), equity in European banks (Credit Suisse) and in Asian oil companies, investment in European football clubs (Paris Saint Germain) as well as farmland across the world. The goals of these ventures are partly economic diversification but they are also politically motivated.

Qatar's foreign policy has been described as deeply pragmatic (Khatib, 2013) to support its geopolitical aims according to useful ventures. Qatar cannot be understood without emphasizing its close links with political Islamist movements such as the Muslim Brotherhood across the region. The intellectual leader of the Muslim Brotherhood movement, Sheikh Yusuf al-Qaradawi, has been granted residence by Qatar since 1961. He cultivated the Muslim

Brotherhood movement across the region. Qatar has also supported the Islamist opposition groups in Libya and Syria, the former Egyptian government under President Morsi, the Palestinian Hamas and, most importantly, the National Congress Party of President Al Bashir in Sudan. Qatar's 'web of influence' has been widely extended to parties that share the same political beliefs (Achcar, 2013; Hammond, 2013; Khatib, 2013).

It is therefore no surprise that Qatar has also been actively involved in Sudan. Qatar has brokered an agreement in Darfur by hosting a peace conference in Doha in 2011 (Bartlett, 2013). All of this diplomacy and engagement tells a certain story about the foreign policy approach of Qatar over the past years. It is driven by the desire to gain geopolitical influence across the world and especially in the Arab world. Translated into the expansion plans of the range of choice of farmland investment in Abu Hamad, the 'web of influence' expresses and exemplifies the nature of Qatar's foreign policy paradigm that has been driven by a desire to influence the wider politics of the Arab world. The impact of Qatari activities in Sudan on the water security of downstream Egypt is just one such example.

Qatar's proposed 1bcm per year diversion of the Nile would have provided Qatar with increased leverage in Nile Basin water hydropolitics. Translated into a percentage, the 1bcm per year would have meant that approximately 1.2 per cent of Nile waters would have been used to irrigate food in Abu Hamad. In other words, Qatar would have been able to control 1.2 per cent of water in a highly politicized basin. This could have been used as a political bargaining chip in foreign policy. However, Qatar's role has declined dramatically since 2013. Qatar's persistent support for the Muslim Brotherhood movement across the region angered other Gulf states, namely Saudi Arabia and the United Arab Emirates (UAE) (Dickinson, 2014). Saudi Arabia and the UAE even withdrew their ambassadors from Qatar for eight months from March to November 2014 to signal discontent about Qatar's geopolitical ambitions. Relations have since improved after Qatar agreed to significantly decrease support for Muslim Brotherhood parties and movements in the Arab world (Aboudi *et al.*, 2014).

Declining support for the Muslim Brotherhood movement has also affected the agricultural investment of Qatar in Sudan. Qatar's interest in Sudanese faded when the old leadership was replaced by the former crown prince on 26 June 2013. Hassad Food has since then actively invested in Australia, Latin America and Pakistan but has not made any attempt to further tap into the political 'breadbasket' in Sudan.

Conclusions

This chapter has shed light on a largely untold story of land investments in the Nile Basin. By presenting external drivers for investment, the legitimate reasons for alternatives to food trade dependency were presented. However, any analysis of drivers is incomplete without taking into account the domestic political drivers. In Qatar, these drivers stem from shadow state politics that emerge from informal networks around the respective leaders. Agriculture and food security were part of an internal political power struggle between a leadership bracing itself for the future of the State of Qatar. While one political elite sought to use environmental initiatives as a way to gain political influence, the other side intended to use agricultural investment as a way to expand a 'web of influence' in the Arab world. However, disputes with other Gulf states led to the gradual demise of investment plans and their implementation.

Qatar's story shows that food is decisively political. It also shows that the term 'land grabbing' is an inaccurate description for highly sophisticated political processes stemming from social history and political bondage. 'Land grabs' or inward investment in land are part of food politics, which is still an under-researched subject in international political economy. Transboundary water politics are also subordinate to food politics, which are defined by deeply rooted political interests that have access to capital from the world's largest companies and sovereign wealth funds (SWFs). Understanding food politics will be crucial to achieving food and water security in the Nile Basin in the coming decades.

Notes

1 Personal discussion with former representatives of Hassad Food, August 2015.
2 Virtual water is defined as 'the volume of freshwater used to produce the product, measured at the place where the product was actually produced' (Chapagain and Tickner, 2012).
3 Interview with Fahad Al-Attiyah (QNSFP), May 2011.
4 Interview with Fahad Al-Attiyah (QNSFP), May 2011.
5 Interviews with a representative of Hassad Food May 2011; interview with Fahad Al-Attiyah (QNSFP), May 2011.
6 Interview with Carter Coleman, Agrica Rice, December 2012.
7 Interview with Dr Turki Al-Faisal Al-Rasheed (Owner of Golden Grass Inc.), November 2012.
8 Interview with a representative of Citadel Capital, May 2012; interview with a representative of Hassad Food May 2011; interview with Fahad Al-Attiyah (QNSFP), May 2011.
9 Email conversation with a journalist of an international newspaper, April 2013.
10 Interview with Fahad Al-Attiyah, May 2011.
11 Interviews with a representative of Hassad Food, May 2011.
12 Email conversation with a journalist of an international newspaper, April 2013.
13 Interview with Fahad Al-Attiyah, QNSFP, May 2011
14 Interview with a former advisor to the QNSFP, June 2013.
15 Interview with George Malbrunot, *Le Figaro*, May 2013.
16 Interviews with a representative of Hassad Food, May 2011 and December 2012.
17 Interview with Dr Bayoumi Attia, former chief adviser to the Ministry of Water and Irrigation in Egypt, April 2012.

References

Aboudi, S., Abdelaty, A. and Bakr, A. (2014). Saudi Arabia, UAE and Bahrain end rift with Qatar, return ambassadors. *Reuters*. 16 November. Retrieved from www.reuters.com/article/2014/11/16/us-gulf-summit-ambassadors-idUSKCN0J00Y420141116.

Achcar, G. (2013). *The People Want: A Radical Exploration of the Arab Uprising*. London: Saqi.

Allan, J. A. (2002). *The Middle East Water Question: Hydropolitics and the Global Economy*. London and New York: I.B. Tauris.

Anonymous (2013). President Morsi calls for Egyptian 'unity' in face of threats to Nile water. *Al Ahram Online*, 11 June 2013.

Anseeuw, W., Cotula, L. and Taylor, M. (2013). Expectations and implications of the rush for land: Understanding opportunities and risks at stake in Africa. In *Handbook of Land and Water Grabs: Foreign Direct Investment and Food and Water Security*, eds J. A. Allan, M. Keulertz, S. Sojamo and J. Warner, 421–436. Abingdon: Routledge.

Bartlett, A. (2013). The dirty hand of Qatar in Sudan's conflicts. *Sudan Tribune*, 20 June.

Brautigam, D. (2013). Chinese engagement in African agriculture: Fiction and fact. In *Handbook of Land and Water Grabs: Foreign Direct Investment and Food and Water Security*, eds J. A. Allan, M. Keulertz, S. Sojamo and J. Warner, 91–104. Abingdon: Routledge.

Carmody, P. (2012). A global enclosure. The geo-logics of Indian agro-investments in Africa. In *Handbook of Land and Water Grabs in Africa: Foreign Direct Investments and Food and Water Security*, eds T. Allan, J. Warner, S. Sojamo and M. Keulertz, 118–131. London and New York: Routledge.

Chapagain, A. K. and Tickner, D. (2012). Water footprint: Help or hindrance? *Water Alternatives*, 5(3): 563–581.

CIA Factbook. (2013). Qatar. Langley, VA: CIA. Retrieved from www.cia.gov/library/publications/the-world-factbook/geos/qa.html.

COMTRADE. (2013). COMTRADE Database. New York: United Nations Commodity Trade Statistics Database. Retrieved from http://comtrade.un.org.

Cotula, L. (2011). *Land deals in Africa: What is in the contracts?* London: International Institute for Environment and Development.

Cotula, L. (2012). The international political economy of the global land rush: A critical appraisal of trends, scale, geography and drivers. *Journal of Peasant Studies*, 39(3–4).

Crystal, J. (1990). *Oil and Politics in the Gulf: Rulers and Merchants in Kuwait and Qatar*. Cambridge and New York: Cambridge University Press.

Deng, D. K. (2011). *The New Frontier: A Baseline Survey of Large-Scale Land-Based Investment in Southern Sudan*. Norwegian People's Aid, Generation Agency for Development and Transfromation-Pentagon (GADET Pentagon), South Sudan Law Society (SSLS).

Dickinson, E. (2014). How Qatar lost the Middle East. *Foreign Policy*, 5 March.

Embassy of the United States in Khartoum Sudan (2015). *Sudan Sanctions*. Khartoum: US Embassy.

FAOSTAT. (2013). Rome: FAO. Retrieved from http://faostat.fao.org.

Fromherz, A. (2012). *Qatar: A Modern History*. Georgetown: Georgetown University Press.

Hammond, A. (2013). The controversial Qatar–Brotherhood alliance. *The Daily Star*, 29 June.

Hanieh, A. (2011). *Capitalism and Class in the Gulf Arab States*. New York: Palgrave Macmillan.

Hendrickson, M., Wilkinson, J., Heffernan, W. and Gronski, R. (2008). *The Global Food System and Nodes of Power*. Boston, MA: Oxfam America.

International Monetary Fund (2015). *World Economic Outlook Database*, April. Washington, DC: IMF.

Kerr, S. (2013). Fall of Egypt's Mohamed Morsi is blow to Qatari leadership. *The Financial Times*, 3 July.

Keulertz, M. and Woertz, E. (2015). State actors in international agro-investments: The role of China, Russia and Gulf. *Development Policy*, 6.

Khatib, L. (2013). Qatar's foreign policy: The limits of pragmatism. *International Affairs*, 89.

Kinnimont, J. (2013). Qatar's delicate balancing act. *BBC World News*, 16 January.

Land Matrix. (2016). The Land Matrix Database. Rome: Land Matrix. Retrieved from www.landmatrix.org.

Luttrell, C. B. (1981). *A Bushel of Wheat for a Barrel of Oil: Can We Offset OPEC's Gains with a Grain Cartel?* St. Louis: Federal Reserve Bank of St. Louis.

Ministry of Water and Irrigation of the Arab Republic of Egypt (2011). National Water Resources Plan Coordination Project, ed. Ministry of Water and Irrigation. Cairo: Government of Egypt.

Mulligan, M. (2013). The water resources implications for and of FDI projects in Africa: A biophysical analysis of opportunity and risk. In *Handbook of Land and Water Grabs: Foreign Direct Investment and Food and Water Security*, eds J. A. Allan, M. Keulertz, S. Sojamo and J. Warne, 384–406. Abingdon: Routledge.

Murphy, S., Burch, D. and Clapp, J. (2012). Cereal secrets: The world's largest grain traders and global agriculture. Oxford: Oxford Research Reports.

Paarlberg, R. (1980). Lessons of the grain embargo. *Foreign Affairs*, 59, 144–162.

Paarlberg, R. (2010). *Food Politics: What Everyone Needs to Know*. Oxford: Oxford University Press.

Piesse, J. and Thirtle, C. (2009). Three bubbles and a panic: An explanatory review of recent food commodity price events. *Food Policy*, 34, 119–129.

QNSFP. (2013). Qatar National Food Security Programme. Retrieved from www.qnsfp.org (offline).

Sojamo, S., Keulertz, M., Warner, J. and Allan, J. A. (2012). Virtual water hegemony: The role of agribusiness in global water governance. *Water International*, 37, 169–182.

SWF (2013). *Qatar Investment Authority*. Las Vegas, NV: Sovereign Wealth Fund Institute.

Teubal, M. (1993). Agroindustrial modernization and globalization: Towards a new world food regime. *Institute of Social Studies*, Working Paper Series No. 162.

The Economist (2012). How to feed a planet – continued. *The Economist*. 28 May. Retrieved from www.economist.com/blogs/feastandfamine/2012/05/food.

Tripp, C. (2007). *A History of Iraq*. Cambridge: Cambridge University Press.

Verhoeven, H. (2013). Sudan and its agricultural revival: A regional breadbasket at last or another mirage in the desert? In *Handbook of African Land- and Watergrabs*, eds T. Allan, J. Warner, S. Sojamo and M. Keulertz, 41–54. London and New York: Routledge.

Water Footprint Network. (2012). Water Product Footprints. Delft: WFN.

Waterbury, J. (2002). *The Nile Basin: National Determinants of Collective Action*. New Haven, CT: Yale University Press.

Woertz, E. (2013a). The governance of gulf agro-investments. *Globalizations*, 10, 87–104.

Woertz, E. (2013b). *Oil for Food: The Global Food Crisis and the Middle East*. Oxford: Oxford University Press.

Woodhouse, P. and Ganho, S. (2011). Is water the hidden agenda of agricultural land acquisition in sub-Saharan Africa? *International Conference on Global Land Grabbing*, ed. Brighton: L. D. P. Initiative.

6 Sudan, 'kingmaker' in a new Nile hydropolitics

Negotiating water and hydraulic infrastructure to expand large-scale irrigation

Ana Elisa Cascão and Alan Nicol

Introduction: Sudan's agriculture is transboundary in nature

Sudan is often presented as the Nile riparian state with the highest agricultural potential, including contributing substantially to food security in the Nile region and beyond. This potential (or perception of) arises out of Sudan's large tracts of fertile land with high potential for irrigation and rainfed agriculture. Sudan's arable land is estimated to be 105 million hectares, of which only around 18 million hectares are currently under cultivation, most of which is rainfed. The current reported irrigated area ranges from 1.2 million to 2.2 million hectares, which is only around 1 per cent of total arable land (Conniff *et al.*, 2012; Johnston, 2012; Arjoon *et al.*, 2014). Full harnessing of that potential is complex and in the past has been affected by a protracted civil war and generalized political instability, economic and political isolation, incoherent water and agricultural policies, and low levels of public and external investment, among other factors. The main constraint on the achievement of Sudan's full agricultural potential lies in the need for future development of water resources infrastructure to enable large-scale irrigation (LSI) expansion.

In Sudan, the largest potential for irrigation development can be found in the Blue Nile Basin. Considering that the Blue Nile is a shared river and alone contributes nearly two-thirds of the average annual Nile flow, the development of appropriate hydraulic infrastructures will involve complex international political negotiations with its riparian neighbours. The Nile is a highly politicized river basin, and Sudan's potential for large withdrawals for irrigation expansion is intimately embedded in Sudan's complex relationship both to Egypt and Ethiopia. On the one hand, Sudan's expansion of hydraulic and agriculture infrastructure is limited by the country's commitment to existing water agreements, in particular the 1959 bilateral Nile Waters Agreement (NWA) with Egypt. On the other, Sudan is the country that has the greatest potential to influence the flow of Nile waters to its neighbour, due to the fact that it has far more available land to irrigate than other upstream riparians, which means the greatest potential for consumption of water.

This chapter discusses exactly how Sudan has the greatest potential to influence the hydropolitical relations between Nile riparian states. As Sudan emerges from a long-standing dyadic relationship dominated by the NWA, it finds itself in a powerful new role as 'kingmaker', able to influence the Nile development outcomes of both Ethiopia and Egypt. In this 'kingmaker position' (in game theory, a kingmaker is a player who lacks sufficient resources or position to win at a given game, but possesses enough resources to decide which of the other players will eventually 'win'), Sudan can choose to align either upstream or downstream, having in mind its own interests and acting in a more autonomous fashion than hitherto. The *first section* of the chapter provides a brief historical account of how Sudan in the past has negotiated in order to initiate its post-independence 'hydraulic mission'. The *second and third sections* look at the new hydropolitical dynamics emerging in the past decade, which are driven increasingly by the changing nature of the Sudanese political economy and new relations with its key neighbours. The *last section* analyses how current and future changes in Sudan's positionality on the Nile signify a structural shift in broader transboundary relations within the wider Nile Basin hydropolitical landscape.

Sudan's agricultural ventures in the post-independence period: rise, crises and fall

Twentieth century: Sudan's 'hydraulic mission' and agriculture ambitions

The history of Sudan as a state and as a political economy is intimately linked to the development of the Nile waters, including infrastructure development and agricultural expansion. In the pre-independence period, the building of the Sennar Dam (1926) on the Blue Nile enabled development of the Gezira Scheme which at the time was the largest single irrigation scheme in Africa. The purpose of the Gezira Scheme was largely to service the demand for cotton within the British Empire at a time where there was widespread recognition of Sudan's immense agricultural potential (Knott and Hewitt, 1994; Tvedt, 2004). The 1929 Agreement (between Egypt and the United Kingdom on behalf of Sudan) established only limited water quotas for Sudan – 4bcm (billion cubic meres)/year, against 48bcm for Egypt. Further expansion of Gezira's command area was dependent on the construction of new water infrastructure beyond Sennar, with larger storage capacity and involving a new set of negotiations (Waterbury, 1979).

On the eve of Sudan's independence in 1956 (from the Anglo-Egyptian Condominium), negotiations between Egypt and Sudan began in earnest and culminated in the 1959 NWA, which included construction of the High Aswan Dam in Egypt. For Sudan, the NWA was an attempt at increasing the 1929 quota and getting 'permission' to build more storage infrastructure to expand agriculture upstream. For Egypt, the NWA helped to constrain

Sudan's future utilization of Nile waters, by locking it into a formal agreement and establishing the Permanent Joint Technical Committee to assess and monitor future utilization of Sudan's quota (Agreement, 1959, Art. 4). As an outcome of the 1959 negotiations, Sudan tried to incorporate the construction of storage dams in order to expand land under cultivation and establish new irrigation schemes. On the Blue Nile, as part of the agreement Sudan was allowed to develop the Roseires Dam (around 3bcm of storage) and subsequently to extend the Gezira Scheme up to around 800,000 hectares of land (Conniff *et al.*, 2012). Sudan could also construct the Khashm El-Girba Dam (around 1bcm of storage) on the Atbara River and establish the 'New Halfa' irrigation scheme to compensate Sudanese Nubians displaced by the Aswan High Dam. The NWA set a new maximum water quota of 18.5bcm/year for Sudan, which in practical terms came to define the country's future irrigation limits (Agreement, 1959, Art. 2.2). However, due to the limited total storage capacity of its dams, it was not anticipated that Sudan would be able to fully utilize its water quota in the ensuing decades. Only with greater regulation of the Blue Nile through dam construction and heightening since the mid-2000s has this possibility now become a reality, with potentially significant consequences for Egypt's water availability and signifying an end to the status quo ante.

In the mid-1970s, an awakening pan-Arabism and demand for food in the Arab world made Sudan attractive to regional investors, in particular from the water-scarce Arab Gulf countries. The oil boom in the Gulf countries after the oil embargo in 1974 spurred belief in Sudan as a potential 'breadbasket' with petro-dollar investments in major agricultural expansion encouraged during the Nimeiri regime. However, debt crises in the 1980s and political turmoil in the Gulf challenged many assumptions underlying the Arab region 'breadbasket' idea, including availability of investment resources, indicating that land and water availability alone would be insufficient to develop and sustain agricultural development in Sudan (Keulertz, 2012; Verhoeven, 2012; and see Chapters 4 and 5 in this volume). As a result, high expectations that Sudan would be transformed into a 'breadbasket' gave way to little actual development of the country's agriculture.

Up to the end of the 1980s, Sudan's stance on Nile hydropolitics was shaped by its need for hydraulic infrastructure and agricultural development in accordance with the provisions of the 1959 Agreement. Water withdrawals for irrigation expansion (in the Blue and White Niles and the Atbara River) had never exceeded the quotas as defined in the agreement. Sudan generally maintained its hydraulic alliance with Egypt even when bilateral relations between the two countries were strained (e.g. after the 1989 coup when the Islamist-led government of Omar el-Bashir came to power) (Waterbury, 2002; also cf. Chapter 4 in this volume). Sudan has rarely publicly contested the 1959 Agreement but rather contributed to cementing Egypt's hydro-hegemonic position within the whole basin through the NWA 'dyad' between the two states in the river system. Historically, this has 'locked in' all the Nile waters, a fact on the ground that began during a colonial period and still surfaces during tensions between Egypt, Sudan and other upstream riparians, but most notably with Ethiopia (Cascão, 2008, 2009). The above factors together form the backbone of the long-standing 'status quo' situation, referred to frequently in this chapter. More recently, however, this status quo has shifted in response to a new power dynamic rooted in upstream actions – including hydraulic construction – that has led to a series of new 'facts on the ground' (Nicol and Cascão, 2011). To understand why and how Sudan's position has changed, we now need to understand the fundamental shifts that have taken place within Sudan in the past two decades.

The 1990s: oil and energy agendas predominance over agriculture

The 1990s brought about key political and economic change to Sudan with impacts upon national agricultural and energy agendas. The changes included the reigniting of war in the south amid continued Islamization in the north following the 1989 coup, and related effort by Khartoum to develop – using Chinese expertise – Sudan's key oil reserves, much of which straddles the north–south border (Collins, 2008). The prize of crude oil exploitation after Chinese completion of the pipeline to Port Sudan provided Sudan's government with new financial levers with which to develop major infrastructure, including transport and energy-related projects. However, during the same period, under a rapid policy of economic liberalization, the government reduced its engagement in agriculture and introduced a more commercial model, the end result being a waning of the importance of agriculture in terms of economic output, public investment and position within the wider political agenda (Mahgoub, 2014). Oil exploitation had, in effect, contributed to a strategic shift away from the agriculture (Stads and El-Siddig, 2010).

In hydropolitical terms, neglecting agriculture led to the degradation of water infrastructure (mainly irrigation canals and schemes), and less water being consumed in the Gezira and Blue Nile schemes. Generally it was estimated that Sudan was using around 14–14.5bcm of Nile water per year; but more recent figures suggest that Sudan had been using less than that – around 12–13bcm/year (MAF, 2000; Mohieldeen, 2007; Salman, 2011). For Egypt, this allowed a water 'dividend' to flow annually (some estimate as much as 4–5bcm/year) into Lake Nasser – or referred to by some Sudanese authors as a 'water loan' (El-Zain, 2007). This allowed Egypt to increase withdrawals beyond the 55.5bcm year on year leading to substantial extension of irrigated agriculture in Egypt during the 1990s, including new irrigation projects in the Nile Delta and the controversial Toshka 'Southern Valley' schemes. This out-of-basin project located southwest of Lake Nasser was initiated in the mid-1990s and became operational in 1997 when the Toshka Diversion Canal was inaugurated. The scheme effectively 'locked in' surplus flows so that Egypt could fully utilize not only Sudan's historical 'water loan' but also the exceptional water flows of the 1990s (Waterbury, 2002). The Toshka remains controversial among neighbouring riparians because it suggests a strengthening of Egypt's claim in the event of any future upstream increases in water consumption – including in Sudan's agriculture sector – based on the principle of prior use (Warner, 2013).

The 1990s also witnessed a new trend in Sudan: the interest in the Nile waters shifting from an agricultural orientation towards a hydropower agenda, with new financing available from oil revenues at a time of historically high oil prices. In earlier decades, Sudan had been mainly interested in developing hydraulic infrastructure for irrigation. The new oil economy enabled the country to respond to increasing demand for electricity in Khartoum and other expanding cities, as well as demand from privately owned pump schemes on the White Nile, through developing hydropower. Roseires and the other dams had limited capacity to produce energy because of their original design specification for irrigation water storage (hydropower was mainly regarded as a 'by-product'), and decades of siltation had substantially decreased their storage capacity for irrigation and hydropower (Awulachew *et al.*, 2008). The Roseires Dam (before heightening) had around 280 megawatts (MW) of installed capacity, and the Khashm-El-Girba, Sennar and Jebel Aulia dams together around 60MW. It was against this backdrop that the Merowe Dam project, for which planning started in the late 1990s and construction lasted between 2003 and 2009 at a cost of US$1.2 billion, provided additional generating capacity of 1,250MW, and increased threefold Sudan's national energy supply (Liu *et al.*, 2013). It was only possible to embark on this expensive project because of Sudan's strengthening diplomatic relations with China and key Arab Gulf states, which provided the foundations to finance and construct such mega projects. For Sudan, the Merowe project both provided much-needed energy and at the same time also became a powerful political statement by the Islamist government, illustrating a vision of 'national development' and economic modernization under a new nation-building project (Verhoeven, 2012).

In regional hydropolitical terms, Sudan's unilateral decision to build the Merowe Dam could not be publically contested by Egypt because it lay within the provisions of the 1959 Agreement, as it was still within Sudan's quota of 18.5bcm and the agreement did not prevent Sudan from building further infrastructure. However, the decision rang alarm bells with Egypt's authorities and exposed Egypt's vulnerability in the long-term to Sudan's willingness and ability to build up its water infrastructure without Egypt being able, or allowed, to control the process (Tvedt, 2013). By taking the decision unilaterally, Sudan signalled change in the midstream country's hydropolitical thinking. The Sudanese government's decision to 'transfer' the equipment, workers, and financial and technical support directly to the Roseires site soon after construction ended at Merowe indicated Sudan's new level of ambition to utilize fully its existing quota and to manage the Blue Nile in closer alignment with its upstream Ethiopian neighbour (Cascão, 2009).

The 'new' Sudan in the twenty-first century: changing approaches to the Nile waters

Twenty-first century Sudan: changing political economy and back to the 'green gold'

Sudan began the twenty-first century in transition, both politically and economically. The first decade involved profound political and economic change, including separation of the south and the loss of key oil revenues to the new Government of South Sudan, as result of the Machakos Peace Accords. This eventual loss of significant oil revenues triggered a wider and more significant political-economic shift away from reliance on oil revenues to a realization that broader development of agriculture in the north – and, in particular, LSI – would have to substantially contribute to Sudan's future economic development. This new direction reflected a wider global political economy that had shifted significantly after the 2008 food price spike, and had placed a spotlight on countries such as Sudan with major land and water resources at their disposal (cf. Chapter 3 in this volume).

The government in the north reprised an earlier focus on agriculture both in order to reduce its own rising food import costs and to increase its presence in global commodity markets, particularly for sugar. Recognizing the challenges of reliance on oil prices (even after loss of revenues Khartoum would still receive pipeline transit fees) and the need for major reform in the agricultural sector in order to accede to the World Trade Organization, Sudan launched in 2006 its Green Mobilization Programme and in 2008 the Agricultural Revival Programme, both of which aimed at revitalizing the agricultural sector. The Revival Programme focused on increasing agricultural development and export earnings with a strong emphasis on sugarcane (RoS, 2008). Sudan aspires to become one of the top five global producers of sugarcane, or so-called 'green gold' (Nazir and Desai, 2001). Representing a key nexus between new agriculture schemes, the commercial goals of private companies and investors within a more liberalized economy, as well as the government's urgent need to generate foreign currency, this 'green gold' vision involves increasing production to 10 million metric tonnes (MT) by 2020, bringing it into the top three African producers after Egypt and South Africa. At the same time, the government has similarly bold ambitions to substantially expand the area under cotton and wheat production by doubling wheat production and nearly quadrupling cotton production from 2007/2008 levels (RoS, 2008).

To drive this production shift, the government has encouraged new types of business models to emerge combining privately driven investment under mixed funding (e.g. Kenana Sugar Company [KSC]) and state-owned companies (e.g. White Nile Sugar Company [WNSC]). The 2008 Green Mobilization Programme focused, in particular, on increasing the area under sugarcane production in the White Nile State and increasing the area under cultivation in the Blue Nile State and the southern part of the Gezira Scheme after heightening of the Roseires Dam. At the same time, the plan envisaged a more integrated production chain, which would utilize by-products more effectively for bio-energy, animal feeds as well as food products.

This new agricultural policy in Sudan has critical hydro dimensions to it. First, with envisaged substantial expansion in the irrigated area, including the production of highly water-consumptive crops, such as sugarcane, the question is whether existing infrastructure constraints can deliver the required amounts of water at the right time and, if not, what does this mean for new alliances with upstream Ethiopia in the construction and operation of future water-related infrastructures? Second, given that water for irrigation to a large extent represents a net loss to the Nile River system through evapotranspiration, how will this impact upon the 1959 NWA? These questions pose new challenges for the existing Egyptian–Sudanese dyad, and, ultimately, raise questions whether or not the status quo can remain, particularly in the context of a newly assertive upstream Ethiopia.

Cooperation in the Eastern Nile: Sudan's strategy to get more energy and water

As analysed previously, from the late 1990s onwards Sudan used oil revenues and its strategic partnerships with China and Gulf countries to leverage funding and political influence to initiate a discreet (and largely) unilateral strategy of developing the Nile waters. The first of those projects was the large-scale hydropower Merowe Dam, but there had been many in the pipeline (Cascão, 2009). Although unilateral, these plans have not prevented Sudan from additionally seeking to use multilateral cooperation to achieve its goals. From the mid-1990s, and particularly since its institutionalization in 1999, Sudan has engaged effectively in Nile cooperation under the Nile Basin Initiative (NBI). The stated goal of the NBI is 'achieving sustainable socio-economic development through equitable utilisation of, and benefit from, the common Nile water resources' (NBI, 2012, 2014). Sudan has always been one of the riparians more interested in multilateral cooperation and in specific investment projects developed in particular under the NBI's Eastern Nile Subsidiary Action Programme (ENSAP). Potential benefits for Sudan from upstream investment projects are considered multiple and are detailed in Sudan's NBI country paper (NBI, 2012). Below, we present an analysis of the most significant ENSAP projects in order to identify the specific benefits likely to accrue.

ENSAP/ENTRO projects: benefits for Sudan

Under the NBI eight different ENSAP projects were initially identified by the three Eastern Nile countries (Ethiopia, Sudan and Egypt). The underlying assumption was that the projects would bring benefits for all three countries and that all identified projects should be jointly planned, prepared and later implemented (Cascão, 2012). ENTRO (the Eastern Nile Technical Regional Office) would be the technical office in charge of delivering the projects. It is not the goal of this chapter to go into the details of each project, but rather to highlight the envisaged benefits to Sudan.

Knowledge/planning benefits

ENSAP projects such as the fast-track Eastern Nile Planning Model, Flood Preparedness and Early Warning, and Watershed Management projects have contributed to knowledge of the Sudanese authorities in several critical areas. Floods that are frequently a threat to communities along the Blue Nile in Sudan can now more effectively be handled through data-sharing with Ethiopia (ENSAP, 2008, 2009). Decades of poor watershed management in upstream catchments in Ethiopia have led to high silt loads with additional costs incurred in Sudan, but now cooperation with Ethiopia is expected to substantially reduce siltation downstream. The Planning Model has contributed to understanding the benefits that Sudan can receive from joint planning and coordination with Ethiopia (ENSAP, 2008, 2009).

Energy benefits

Two specific ENSAP projects, the Regional Power Trade Study and Ethiopia–Sudan Transmission Line, have also provided important potential benefits. A transmission line now connects the national power grid of Sudan with Ethiopia, the future powerhouse in the region, enabling cross-border power trade (ENSAP, 2008, 2009). For Sudan, this power connection entails more available electricity (and at an anticipated cheaper price) in support of agricultural development (water pumping, irrigation pumping, etc.). The Regional Power study (ENSAP, 2008, 2009) highlights the greater benefits of hydropower development in the event that Ethiopia, as the main producer and exporter of energy, and Sudan, as an important player within the regional 'energy corridor', are connected to a wider grid linking to Egypt, the Middle East and potentially onwards to Europe.

Development in the Blue Nile Basin

Interlinked with the other ENSAP projects but larger in terms of ambition is the Joint Multipurpose Project (JMP). The JMP was designed as an inclusive approach to the management and development of all three major river systems comprising the Eastern Nile Basin (Blue Nile, Atbara/Tekezze and Baro-Akobo Sobat). The JMP was suggested to provide a basket of benefits including energy, agriculture, water storage, joint planning and trade for all four riparian countries. The 2008 JMP Scoping Study identified the Blue Nile as the basin with the highest potential to generate benefits for all countries (Blackmore and Whittington, 2008). This study included an analysis of different scenarios and concluded (along the lines of other modelling studies) that upstream infrastructure in the Blue Nile Basin can provide a number of direct and indirect benefits to downstream riparians (Guariso and Whittington, 1987; Block *et al.*, 2007; McCartney *et al.*, 2012; Satti *et al.*, 2015). The JMP study detailed many of the positives that were projected to benefit Sudan directly. These included:

- The construction of large-scale dams on Ethiopia's portion of the Blue Nile considerably increasing hydropower production and, in theory, providing access to hydropower at lower prices for neighbouring states through favourable power purchase agreements.
- Increased large-scale water storage capacity that could promote water conservation – through reduced evapotranspiration at higher altitudes and in reservoirs with a more favourable surface-to-volume ratio – thereby making more water available for irrigation development downstream, in particular within Sudan's portion of the Blue Nile Basin.
- Upstream infrastructure is expected to regulate the timing of flows and increase water availability during dry months, which would allow Sudan to irrigate during these months (at present Sudan mainly irrigates in the wet months), strengthening Sudan's agricultural production and output.
- Upstream infrastructure contributes to improved flood and drought control, improved watershed management, reduction of silt/sediment load, and as such decreases Sudan's exposure to hydrological/climate hazards as well as reducing the costs associated with the maintenance of downstream hydraulic infrastructures.

Cooperation in the Eastern Nile: challenging established norms

The 2008 JMP scoping study represents an important milestone for Nile Basin cooperation and should be seen as a landmark. It was the first time that such a study had been jointly commissioned by the three Eastern Nile countries and was expected to contribute to crucial changes in the decision-making process in the basin, effectively breaking a mould that had been established around the 1959 NWA. Up to that point, the pattern had been discussions that followed the tramlines (and confines) of Egypt and Sudan restating their historic 'water rights' as defined in the NWA, that these rights had to be respected and that projects upstream should be avoided in order to prevent infringing such rights. With the JMP project, the expectation was that for the first time Eastern Nile decision-makers could sit at the same table jointly debating and deciding over the best case scenario for the development of the Eastern Nile sub-basins, looking at the opportunities and benefits that joint projects could bring to all. The JMP Scoping Study suggested that the project that would maximize joint benefits would be a cascade of dams on the Blue Nile, maximizing rewards and minimizing risks (Blackmore and Whittington, 2008). High expectations existed at ENTRO, and among its member states, that the multi-billion dollar

JMP project could become the first large-scale project in the Eastern Nile Basin applying a genuine transboundary approach. This would include:

1 planning and implementation based on a regional decision-making processes;
2 regional confidence-building based on joint communication and consultation mechanisms;
3 a benefit-cost sharing formula between the riparians under a 'no-borders perspective'; and that ultimately
4 it would contribute to ensure efficient and optimal use of the Nile waters through equitable and reasonable utilisation.

(Eldaw and Fekade, 2009)

It was anticipated that the JMP was the transboundary project that could unlock further linkages in terms of regionalism and trade, and provide transformational socio-economic benefits to the region (Eldaw and Fekade, 2009).

However, this shift to 'regionalism' as an approach quickly became overshadowed by old-fashioned hydropolitics. Soon after the document was released, the JMP scoping study was 'classified' at the insistence of Egypt, and later the JMP was temporarily closed after the 2010 disagreements between Nile countries involved in the Cooperative Framework Agreement (CFA) negotiations. The decision of the upstream countries to sign the CFA (Agreement, 2010), even against the position of downstream states, resulted in Egypt and Sudan 'freezing' participation in the NBI and ENTRO (ST, 2010). Sudan resumed its membership in 2012 subsequent to which it emphasized a 'midstreamer' perspective including trying to coax Egypt back into the NBI, though so far without success (ST, 2012c; AFP, 2014). This complex hydropolitics contributed to slowing the pace of multilateral investment projects, with large-scale projects such as the JMP unable to progress not least because a lack of consensus made financial and technical support impossible. The failure of 'regionalism' as an approach created incentives for countries to act unilaterally. In April 2011, Ethiopia announced to the region and wider world the decision to construct one of the dams included in the JMP scoping study, but with its own technical and financial resources. Known as the 'Border Dam' under the JMP (the most downstream dam in the envisaged cascade) it would become the Grand Ethiopian Renaissance Dam (GERD) and later in the chapter we analyse how Sudan's reaction to the GERD exemplified its new positionality on the Nile.

In conclusion, the multilateral cooperation process of the NBI/ENTRO contributed to Sudan establishing a new, more independent position within a new hydropolitical landscape. It also highlights Sudan's acknowledgement that greater cooperation and joint infrastructure can actually enhance its own national capacity to develop further its Nile water share. The NBI and the ENSAP forums created neutral space within which Sudan could articulate a shift towards an alliance with other upstream countries, representing a *de facto* break with the long-standing norms entailed in the Egyptian–Sudanese relationship.

Damming the Blue Nile: Roseires and GERD – the possible cascade?

The previous section highlighted the extensive benefits that a joint and multilateral approach to the development of the Blue Nile waters could bring to the Basin, and in particular the specific benefits for Sudan. However, wider geopolitics and the hydropolitics of the Nile itself have caused this optimistic scenario to fade, and the riparian countries have developed individual hydraulic infrastructure. In the shorter term, in the absence of JMP implementation, Sudan has continued to develop other alternative options for energy and irrigation expansion, of which both heightening the Roseires Dam and political support to Ethiopia's construction of the GERD provide evidence. Part of the continued complexity of Nile Basin relations is that Sudan and Ethiopia are *de facto* implementing the concept of the Blue Nile cascade though not under the ENTRO/NBI institutional auspices or perspectives, and for that matter not as part of an officially sanctioned agreement between the parties. In this section we analyse the relevance and regional hydropolitical dimensions of two of these 'national' projects developed in the Blue Nile: the heightening of Sudan's Roseires Dam finalized in 2012, and the GERD, the first large-scale hydropower dam constructed on the Ethiopian Blue Nile, and planned to be completed in 2017.

Heightening the Roseires Dam: an old ambition

> Heightening of Roseires Dam is Sudan's real oil.
>
> (President Al-Bashir; cf. DIU, 2008)

Revival of the irrigation sector to underpin Sudan's agricultural development requires more water storage. Although new irrigation schemes are being planned or initiated on the White Nile and Atbara River, it is the Blue Nile schemes that are a higher priority because of their far greater agricultural potential. However, the extension of existing schemes and planned new schemes (see Table 6.1 and Figure 6.1) requires far more water than had been originally planned for in the old Roseires Dam reservoir, which held around 3bcm. Plans to heighten the Roseires Dam and increase the storage capacity of its reservoir have been discussed since initial construction in the 1960s. The Roseires Dam was originally designed to be heightened and include additional gates and spillways that would connect to additional canals conveying water for irrigation in farther flung irrigation schemes (El-Zain, 2007). In the 1990s, the decision to proceed with the dam's heightening accompanied a generalized policy of substantially expanding the country's level of irrigated agriculture production (Faki *et al.*, 1995). However, according to Sudanese analysts the plans never materialized due to a combination of Sudan's national technical, political and financial constraints and Egypt's opposition, aware that heightening would enable Sudan to substantially increase the utilization of its water quota (Hamad and El-Battahani, 2005; Cascão, 2009).

Table 6.1 Proposed irrigation development for the Blue Nile Basin in Sudan

Project	*Description (see Figure 6.1 for locations)*	*Estimated completion date* *# 2008 (McCartney et al., 2012)* *◊ 2015 by the authors*
Heightening of Roseires Dam	Roseires Dam raised by 10m to provide total storage of 7.4bcm (± additional 4bcm)	# Medium term ◊ Project completed: the project was initiated in 2008 and completed in 2012. Interesting to observe through the literature/reports published prior to 2008 is that the project was not expected to materialise in the short term.
Extension of Rahad Irrigation Scheme	(Current irrigation area: 148,293ha) Additional irrigation area: 19,740ha Rahad II irrigation scheme: 210,000ha Total average annual demand: 2.43bcm	# Medium term ◊ Short/medium term: the Rahad is one of the priority projects, and several national (e.g. KSC) and international investors are interested in developing projects here. Sugarcane (mainly), cotton, groundnuts and sunflower are the main crash-crops. Reports indicate that some leasing contracts have already been signed with investors from Arab/Gulf countries, as well as from Turkey, Brazil and China. Note: the KSC, a public–private joint venture (between the Sudanese government, the Kuwait Investment Authority and the Saudi Arabian government, among other Arab/Gulf investors) is interested in expanding its operations into the Blue Nile region (up to now its schemes have only been located along the White Nile).
Extension of Suki Irrigation scheme	(Current irrigation area: 29,827ha) Additional irrigation area: 2,940/3,361ha Total average demand: 2/2.2bcm	# Medium/long term ◊ Short/medium term: the Suki scheme is often referred to as another priority project, and Sudanese national and local authorities (e.g. Sennar State) are working towards establishing favourable investment frameworks, namely free trade areas. Reports show that the sugar industry (production, factories, etc.) on this scheme is expected to increase substantially. Note: most of the agricultural lands in the Rahad and Suki schemes remain under government control as they are both considered strategic schemes, and are mostly still being managed by parastatal organizations known as agricultural corporations.

Project	*Description (see Figure 6.1 for locations)*	*Estimated completion date* *# 2008 (McCartney et al., 2012)* *◊ 2015 by the authors*
Extension of upstream Sennar	(Current irrigation area: 77,717ha) Additional irrigation area: 39,910ha Total average annual demand: 0.75bcm	# Medium/long term ◊ Reports often mention new investment projects in the Sennar State, although the exact location is not identified and so remains difficult to understand if these projects are within the current schemes or an additional area. The short distance of these schemes to Khartoum is an incentive for investments, due to proximity to transportation (airport, roads to Port Sudan, etc.) for export purposes. Crops being produced include sugar, cereals, oilseeds and fodder. Reported main investors are Saudi Arabia and other Gulf countries. But Sudanese parastatal sugar companies are also expanding in this region.
Extension of downstream Sennar	(Current irrigation area: 167,200ha) Additional irrigation area: 44,110/6,804ha Total average demand: 1.4/1.5bcm	# Medium term ◊ Long-term projects: according to interviews the development of these projects is dependent on the construction of new irrigation canals (main and secondary) on the west bank of the Blue Nile River, though it is mainly linked to the end of civil conflict and political instability in the region (Blue Nile State), ongoing since 2011. The location of schemes very near the Sudan/South Sudanese border also makes these irrigation projects less of a priority when compared to Rahad, Suki or Sennar.
Kenana II and III (new scheme)	Additional irrigation area: 420,093ha Average annual demand: 2.35bcm	
South Dinder (new scheme)	Additional irrigation area: 84,019/48,318ha Average annual demand: 0.54/0.85bcm	# Medium/long term ◊ No specific information found about plans for this scheme.

Source: adapted from McCartney *et al.* (2012) using data collected in 2008; and 2015 updates by the authors.

Additional sources: RoS (n.d.); FAO (2011); SD (2012a, 2013a, 2013b, 2013c, 2013d, 2014); Reuters (2013); FAO AQUASTAT 2015 (and interviews conducted in Sudan – Khartoum, Roseires and Wad Medani – in 2013).

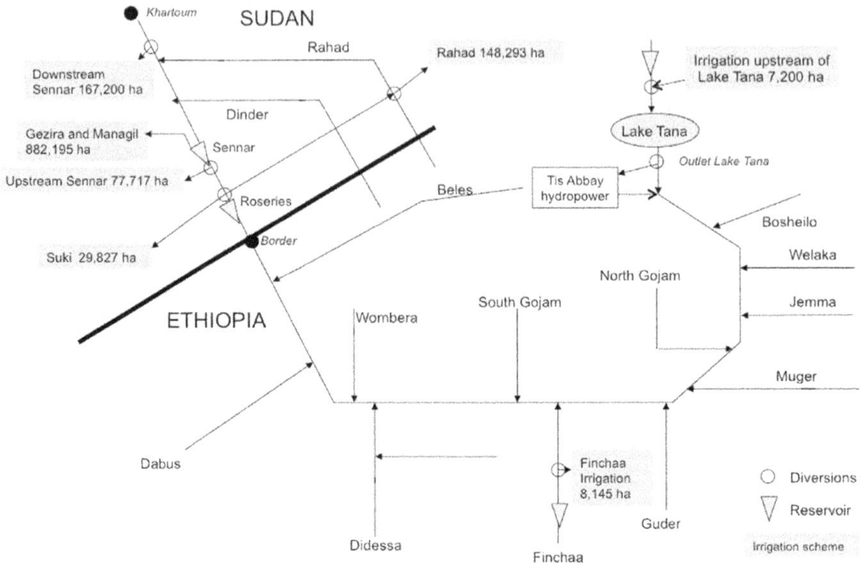

Figure 6.1a Current scenario with existing irrigation schemes and dams (as per 2008)

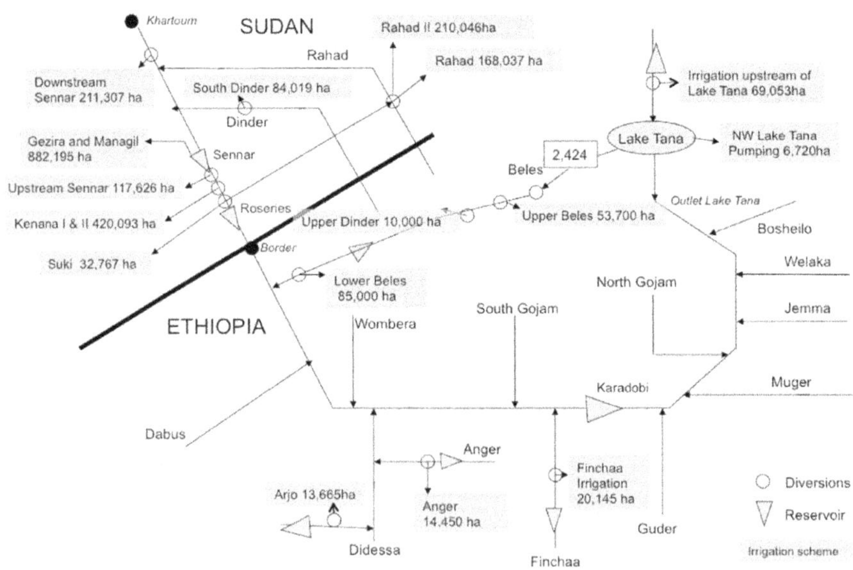

Figure 6.1b Medium-term scenario with proposed dams and irrigation developments (2010–2025)

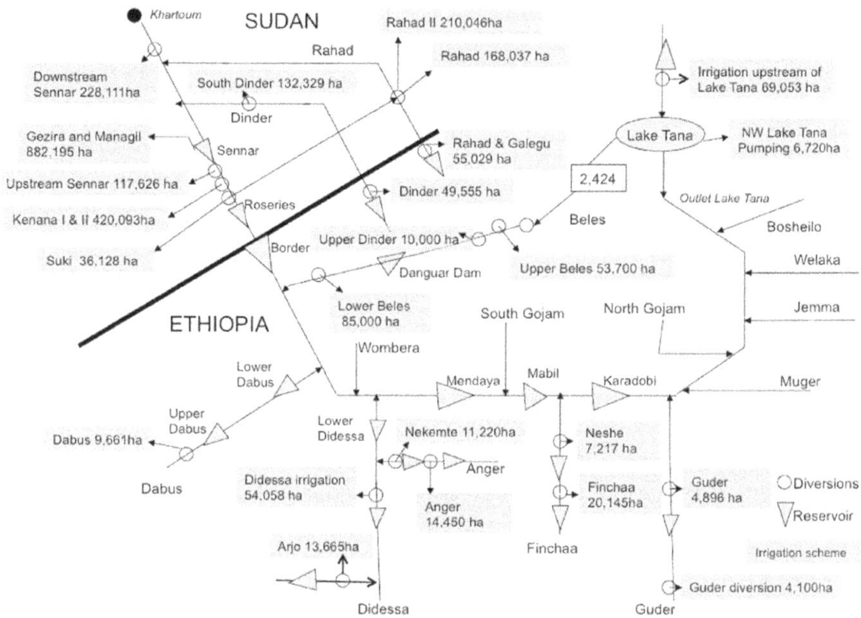

Figure 6.1c Long-term scenario with proposed dams and irrigation developments (2025–2050)

With the renewed interest of Sudan in promoting large-scale agriculture, the plans to heighten the Roseires Dam again became a political priority. This time the works started in 2008, after the project mobilized external funding from Arab partners (Organization of the Petroleum Exporting Countries [OPEC], Arab and Abu Dhabi Funds for Development, etc.) and favourable contracts were agreed with the same Chinese companies involved in construction of the Merowe Dam (ST, 2008). As the earlier statement of President Al-Bashir suggests, the Roseires heightening project has since acquired enormous importance because of its economic significance. This is based on the assumption that expansion of irrigated agriculture, made possible by heightening, has the potential to become a backbone of Sudan's national economy and exports, and a major source of financial revenues (DIU, 2008). The goals of heightening the Roseires Dam by 10 metres were two-fold:

1 to increase the reservoir storage capacity of the dam (the main goal);
2 to increase the hydropower capacity of the dam (a by-product).

The increase of hydropower capacity at the Roserie Dam – from its original 280MW to 400MW (Whittington *et al.*, 2014) – is substantial, but compared to the massive Merowe installed capacity of 1,250MW, it is not that noteworthy. There is little doubt that the most important output of heightening the dam is that it more than doubles the previous reservoir storage capacity – that is, an increase from 3.3bcm/year to 7.4bcm, providing nearly an additional 4bcm of water to expand irrigation schemes in the Blue Nile Basin (SD, 2012b; Reuters, 2013).

Potential new irrigation projects include the extension of existing irrigation schemes such as Rahad, Suki, downstream Sennar and upstream Sennar, but also entirely new schemes such as Rahad II, Kenana I and II, and South Dinder (McCartney *et al.*, 2012). All of these projects have been in the pipeline of the Sudanese government for a long time, but lack of financial resources have prevented their development. Table 6.1 summarizes the size of each project, as well as their water demands. Figure 6.1 shows the locations of the projects. Table 6.1 (third column) provides a brief update on the current status of the main projects, based on fieldwork in the region conducted in 2013, further interviews and media reports (mainly from Sudanese newspapers). However, it is important to mention that is very difficult to get concrete information about who is investing where (which specific scheme), and how much land is actually being leased and for what kind of crops.

Water demands of the Blue Nile irrigation schemes

An important question remains whether or not the additional 4bcm/year after heightening of Roseires provides sufficient water for Sudan to expand all of these irrigation projects in the Blue Nile Basin. The answer appears to be no, based on calculations using the figures provided by McCartney *et al.* (2012). Here we provide a breakdown of the numbers:

- Until recently, Sudan has been irrigating around 1.3 million hectares of land (including the Gezira Scheme) in the Blue Nile sub-basin with water stored behind the old Roseires and Sennar Dams (i.e., around 3bcm/year).
- If Sudan's masterplans for the Blue Nile irrigation schemes – extensions and new schemes (see Figure 6.1 and Table 6.2) – are fully developed, this would add another 0.9 million hectares of land to the existing area under cultivation.
- If Sudan is to expand all these new schemes, calculations show that at least an additional 10bcm of water per year would be needed for full expansion.
- In conclusion, it appears that the additional 4bcm/year of water now stored behind the heightened Roseires Dam would allow Sudan to expand only part of the new schemes.
- Therefore, the full-expansion scenario is only possible if storage infrastructure is developed upstream, underscoring the logic of Sudan's support for the GERD as part of a wider agricultural development strategy.

Table 6.2 Summary of land and water requirements of the Blue Nile irrigation schemes

	Land	*Water*
Current irrigation (with Old Roseires and Sennar storage)	***1.3 million ha of land*** Total area currently under irrigation = Gezira + Managil: 882,000ha + Rahad: 148,293ha + Suki: 29,827ha + upstream Sennar: 77,177ha + downstream Sennar: 167,200ha = 13,305,037ha	***± 3bcm/year*** Total reservoir capacity: Old Roseires Dam: 3.3bcm + Sennar: 0.9bcm = 4.2bcm/year, but significant storage losses (around 85 per cent in Sennar and 35 per cent in Roseires) due to sedimentation
Full-development scenario	***Additional 0.9 million ha of land*** Total area of extended and new schemes = Extension of Rahad I: 19,740ha + Rahad II: 210,000ha + Suki Full Extension: 2,940+3,361ha + extension of upstream Sennar: 39,910ha + extension of downsteam Sennar: 44,110+6,804ha + New Kenanas (II & III): 420,093ha + New South Dinder: 84,019+48,318ha = 879,295ha	***Additional 10bcm/year*** Total average annual water demand of extended and new schemes: Rahad: 2.43bcm + Suki: 2.2bcm + upstream Sennar: 0.75bcm + downstream Sennar: 1.5bcm + Kenana II & III: 2.35bcm + South Dinder: 0.85bcm = 10.08bcm/year

Source: authors.

In brief, the analysis provided above clearly shows how irrigation development in Sudan's Blue Nile Basin has the potential to increase substantially in the medium and long term. However, it is important to keep in mind that this will only be possible if upstream storage infrastructure is in place. The heightening of the Roseires Dam could presage not only the start of a new irrigation era in Sudan, but also signify the end of a long-standing hegemonic status quo in the Nile Basin, as discussed in the next section.

Regional hydropolitical dimensions of the Roseires heightening

Given that the Roseires Dam heightening brings such large benefits to the Sudanese economy, it is pertinent to ask why development has taken so long? Earlier in this chapter, factors were identified including internal conflict(s) and limited financial resources. However, regional factors have also played their part in delaying project implementation. These include the antagonism of Egypt's water authorities to the heightening project, aware that it would enable Sudan to increase its utilization of Nile waters with possible impacts downstream (Waterbury, 2002; Tvedt, 2004). If the heightening 'observes the convention related to the sharing of Nile water signed with Egypt in 1959', as stated by Kamal Ali, the former Sudanese minister of irrigation and water resources (ST, 2008), why then was Egypt so concerned? One needs to understand what exactly the possible impacts are that Egypt is concerned about in order to understand the hydropolitical importance of the Roseires heightening. There are three sets of interrelated consequences:

End of the 'water loan': increased storage capacity behind a heightened Roseires Dam allows Sudan to increase irrigation development and to expand its utilization of the Nile waters (beyond the 12–14bcm the country is currently using); this means a decrease in unused surplus water flowing into Egypt's Lake Nasser.

Egypt's water availability: Egypt will have to become accustomed to limiting its Nile water withdrawals to the 55.5bcm/year quota (as defined by the 1959 Agreement), once the water surplus is no longer available. This is likely to affect water availability for existing irrigation schemes in the Nile Delta, Nile Valley and the Toshka projects, and the country will need to make major adjustments in its current use of water (Whittington *et al.*, 2014), with potentially major economic, social and political consequences within Egypt.

A new hydropolitical threshold: in the likely event that Sudan increases its water utilization to the threshold of 18.5bcm and/or even seeks to exceed that quota, one possible scenario is that Sudan will seek a bilateral renegotiation of the 1959 Agreement to reflect the new water needs/demands. A more extreme – and unlikely – scenario is that Sudan might even question the very existence of the 1959 Agreement and demand a new agreement (possibly a multilateral one with the other Nile riparians).

The historical status quo situation in the Nile Basin has been very much defined by the maintenance of the 1959 Agreement as the single most important element in the hydropolitical landscape, and, historically, a vehicle that served to constrain Sudan's hydraulic aspirations (Eissa, 2008). Sudan's unilateral decision to move forward with heightening of the Roseires Dam (and, prior to that, construction of the Merowe Dam) as well as development of Blue Nile irrigation schemes represents a significant shift away from the status quo though not to the extent of seriously challenging the 1959 NWA. It is, however, a challenge to the notion of a 'unified position' between the two downstream riparians embedded in the agreement and possibly represents a structural change in relations between the two countries under the NWA.

The Grand Ethiopian Renaissance Dam (GERD) and Sudan's positionality

> The GERD is for Sudan what the Aswan High Dam has been for Egypt in the 1970s: an amazing opportunity to increase almost twofold the land under irrigation.

This statement was delivered by a high-level official from the Ministry of Water Resources during a research interview conducted in Khartoum in 2013. It reveals to some extent the Sudanese government's perception of the GERD as an opportunity that could bring a stream of benefits to Sudan, including support to the country's major agricultural ambitions. The statement also partially explains why the highest echelons of the Sudanese government, including President Omar El-Bashir, have been so keen to support the Ethiopian dam project. The following section discusses the direct benefits of the GERD for Sudan's irrigation ambitions, and analyses Sudan's role in the ongoing hydropolitical process and negotiations surrounding the construction and operationalization of the dam.

GERD: the irrigation-related benefits for Sudan

Earlier we presented the benefits that development of upstream dams in the Blue Nile can provide for Sudan. These include flood mitigation, drought management, sediment control and improved navigation. But the primary economic benefits of upstream dams, however, are related to improved flow regulation and increased water storage capacity, enabling substantial expansion of ambitious Blue Nile irrigation schemes. As noted earlier, Sudan needs an additional 10bcm of water per year for full expansion of the Blue Nile schemes, although the heightened Roseires provides an additional 4bcm/year. The key question then is where the additional water requirement will come from. The fact is that Sudan faces a similar problem as Egypt before the Aswan High Dam was built: lack of sufficient over-year storage to supply its irrigation ambitions. Moreover, Sudan has no other sites appropriate for water storage within its national territory for irrigation (Blackmore and Whittington, 2008). In order to intensify and expand irrigation Sudan needs Ethiopia's dam(s), as the required additional water can only be made available if storage infrastructure is built upstream.

Regarding the specific irrigation benefits for Sudan offered by the GERD (or similar infrastructure in the Blue Nile in Ethiopia), several authors have developed different hydro-economic models to show what advantages a large-scale upstream facility can bring (Blackmore and Whittington, 2008; Arjoon *et al.*, 2014; MIT, 2014; Geressu and Harou, 2015; Kahsay *et al.*, 2015; Satti *et al.*, 2015), as summarized below:

- The upstream dam will contribute to increase storage and capacity to regulate the flow, therefore making more water available for irrigation schemes in Sudan.
- This increased water availability will assist Sudan in optimizing irrigation, including its decisions on specific water withdrawals and distribution for improvement and expansion of existing and/or new schemes.
- Significantly, it also provides reliable water supplies during the seasonal low-flow periods, which makes possible or increases summer irrigation.

- An upstream dam can allow Sudan to change the current operating rules of its reservoirs in order to boost irrigation (as well as hydropower) production.
- An upstream dam will provide a more constant flow over the whole year; it will reduce the uncertainty inherent to periods of low flow (seasonal or in case of droughts) and ensure a more continuous irrigation supply and improve water-use efficiency.

In brief, the GERD, through its storage capacity, can provide additional water that Sudan needs to enable development of its Blue Nile irrigation schemes. It is therefore easy to understand why the Sudanese government has been so supportive of the GERD project and its negotiations process. However, collaboration and cooperation between the two countries (as well as Egypt) are required in order to maximize benefits and reduce risks and impacts. It is of particular importance how the countries will agree on joint operation of the infrastructure in the Blue Nile and main Nile and how trade-offs will be negotiated regarding economic sectors (namely, irrigation versus hydropower production). The Sudanese negotiation position is one that highlights the optimization of water availability for national irrigation purposes, particularly in times of low flow. Ultimately, the water to be stored in the GERD will partially determine the success of the current wave of foreign investment in large-scale irrigated agriculture in Sudan because the package offered by the Sudanese government to the investors also included the availability of water.

Hydropolitical implications of Sudan's GERD support

Contrary to Egypt, which mainly sees the GERD as a serious threat to its water security, Sudan considers the GERD to be an opportunity with multiple benefits for the country. This sub-section provides a brief account of the key role Sudan has played in the GERD trilateral process and the active support contributed in the form of public declarations of support, support to bringing the parties together for trilateral talks including high-level diplomatic visits, and the organization of the main trilateral negotiation rounds in Khartoum.

In April 2011, Ethiopian Prime Minister Meles Zenawi made the first announcement that Ethiopia intended to construct the GERD, highlighting the large-scale benefits the dam could bring to Ethiopia and to downstream neighbouring countries. It would be a win–win solution for all, and even Egypt and Sudan could cooperate and eventually contribute financially to the project (*Ethiopian News*, 2011; ST, 2011). The announcement was received with alarm in Cairo and Khartoum; although Sudan's first official reactions were not as negative as in Egypt. Interviews conducted with high-level officials from Ethiopia and Sudan (on several occasions in 2011) indicated that Sudan's support at the highest political level actually preceded the announcement, namely during meetings between Prime Minister Zenawi and President Bashir in 2010. This has never been confirmed officially by Sudan; but years later in December 2014, when President Bashir visited the GERD site for the first time, not only did he confirm his support for the project but he also declared he 'had all the information about the dam since its launch' (ST, 2014c). In mid-2011, news about possible trilateral talks over the GERD surfaced, and the personal involvement of President Bashir in meetings with water authorities of the neighbouring countries assisted the three Eastern Nile countries in setting up a tripartite technical committee, including national and international experts (International Panel of Experts – IPoE). This would assess possible impacts of the dam, foster cooperation and ultimately boost regional development (ST, 2011b). The trilateral technical committee was established and had its first meeting in January 2012. In October, the IPoE was established and started its studies.

Less than one year after the announcement of the GERD, the positive tone of Sudanese officials towards the dam emerged in earnest, which coincided with increasingly strong economic and diplomatic relations with Ethiopia, including several agreements on power connections and trade (ST, 2012a). In March 2012, President Bashir stated explicitly for the first time that Sudan officially supported the GERD, understood the benefits it would bring and would extend all necessary support to ensure completion of the project (ST, 2012b). This official support would be reiterated by President Bashir, his foreign affairs and water resource ministers, several times subsequent to 2012. However, the release of the final report of the IPoE on 31 May 2013 apparently contributed to assuring Sudan about dam safety, and the potential to regulate flows and contribute to expanding irrigated agriculture in Sudan (IPoE, 2013). The release of the IPoE report was followed by numerous official declarations of support by the Sudanese government, downplaying negative impacts and praising the benefits of the GERD for Sudan (ST, 2013a, 2013c). Meanwhile, as relations between Ethiopia and Egypt dwindled after President Morsi's declarations on sabotaging the dam (ST, 2013b), Sudan began to take on the role of mediator between the two conflicting neighbours.

The mediation and diplomatic campaign of Sudan became stronger from this moment, and included regular visits by water and diplomatic experts to Addis Ababa and Cairo, in order to bring countries to agreement on how to proceed with the negotiations and recommendations of the IPoE. Amidst fears in Cairo that Sudan was taking sides, the Sudanese foreign minister insisted on the central role Sudan could play regarding dispute resolution and how it would continue efforts to bridge the gap between the downstream and upstream neighbours (ST, 2014a; *Al-Ahram*, 2014). The Sudanese diplomatic efforts would only be successful after major political change when President Al-Sissi was elected in May 2014 and appeared more open to reactivating the tripartite talks in August 2014. Sudan offered its 'good offices' to host the tripartite meetings and to seek joint cooperation mechanisms, with the midstream country officially being entrusted by both neighbours to play a pivotal role in coordinating all trilateral negotiations (ST, 2014b). In August 2014, the countries resumed the trilateral meetings and would meet six times until the end of December 2015; all except one meeting took place in Khartoum.

Sudan's successful hydropolitical campaign would culminate in the Declaration of Principles for the GERD, signed in Khartoum by the three heads of state in March 2015. The declaration was considered a historical deal, bringing together for the first time the three Eastern Nile countries around guiding principles on cooperative relations. Among these ten principles, several are commonly accepted principles of international water law, such as 'no harm' and 'equitable utilization'; but they also include principles more related to technical issues such as dam security, dam filling (the key challenge), operations policy and the exchange of information (Agreement, 2015). To reaffirm these principles, overcome some challenges and give political and technical drive to the Declaration of Principles, Sudan water and foreign affairs ministers organized new rounds of trilateral meetings in Khartoum in late 2015, and on 28 December the Khartoum Accord was signed by all (ST, 2015a, 2015b). These diplomatic-political outcomes have only been possible because of the active role Sudan has played since mid-2011 in bringing the three neighbours together. After repeated expressions of support to the GERD project since 2011 onwards, President Bashir recently argued during a high-level visit to Cairo that now 'the GERD has become a reality' (*Al-Ahram*, 2015a).

The negotiations and political process around the GERD provide evidence of new hydropolitical realities, with Sudan gaining a key position in the trilateral relations between the countries. Sudan is now more outspoken about its own interests in Nile waters development, and is keener to break away from long-standing bilateral relations with Egypt. Sudan is more supportive of the development of hydraulic infrastructure upstream because of the multiple benefits this can bring to Sudan, particularly if it can assist Sudan's irrigation ambitions. Sudan has actively been reaching out to improve bilateral economic, political and diplomatic relations with Ethiopia, but has also played an important role as mediator by bringing all three countries towards an agreement on the dam. At the end of 2015, the reality is that a large-scale dam on the Ethiopian side of the Blue Nile is becoming a fact on the ground, something that would have been considered almost impossible ten years ago. These new 'realities' represent a fundamental structural change in the Eastern Nile hydropolitical landscape, and herald an era of further change and uncertainty.

Sudan: at the hub of a changing hydropolitical landscape

The Nile basin hydropolitical landscape began the twenty-first century on an optimistic note. After a concerted effort between governments and international development agencies, in 1999 the NBI was established on the back of earlier efforts at cooperation and a decade-long series of basin-wide technical conferences. This new initiative brought together all Nile states at the time and was driven by a vision of cooperative development of the basin's resources to maximize available economic, environmental and social benefits; the assumption was that joint investment projects would be developed to transform the vast potential of water and land resources of the basin into regional energy and food security for all countries (NBI, 2014). This chapter mentioned one of the main projects that could have that transformative potential – the JMP in the Eastern Nile Basin – which, if implemented, could have provided sufficient energy to trade between Africa's Eastern and Equatorial regions.

By the end of the first decade, many of the assumptions made at the start of the NBI and the political pillars underlying those assumptions had been turned upside down: Sudan has been subdivided into two countries and a new Nile state has been added to the riparian list; Egypt has undergone revolution and political turmoil leading to uncertainty regarding its earlier dominant position; and, under the Nile cooperation framework, upstream states have effectively broken away from the basin-wide consensus process under the NBI, forming their own position on cooperation by deciding to ratify the CFA and leaving Sudan and Egypt isolated. Last, but most significantly, Ethiopia has announced the construction of the single largest dam in the Nile Basin since the Aswan High Dam, to be financed unilaterally and outside of NBI framework.

For Sudan, this profoundly new landscape meant considerable challenges and opportunities in its positioning on Nile Basin development. Its response was to generate a new, more powerful role as a midstream-state with 'king-making' capacities. Midstream in both a physical and political sense, Sudan's actions and reactions to the new landscape are having significant consequences for both Egypt downstream and Ethiopia upstream. Sudan is seizing the opportunity to forge a new role, one that is more independent of its earlier 'lock-in' under the 1959 NWA – and with more alliance-building options on the table. But it is also a role steeped in political and economic uncertainty and potential risk, not least of which is its capacity to transform agricultural development in the country. Sudan's political economy has lurched in different directions over the first decade of the twenty-first century, from a major oil producer at a time of rising oil prices, to the loss of oil revenues amid a flat-lining global economy and historically low oil prices. The implications of this lurch to oil and back again is not lost on the government. As a result, Sudan is reawakening its former interest in irrigation expansion tied to major commercial and export ambitions. This renewed interest in LSI poses major questions for Sudan's water requirements, the 1959 NWA water allocations and its relationship to Egypt because of the NWA.

What is now clear is that Sudan's shift back to utilizing its land and water resource potential will rapidly 'use up' any remaining quota that flows to Egypt and, in all likelihood, will start full utilization of any excess flows that may arise in high rainfall years, particularly as upstream infrastructure – including the GERD – makes co-management of the Blue Nile more feasible. Sudan's water requirements therefore might soon reach its NWA quota of 18.5bcm/year, which is used for irrigation schemes in the Blue Nile, White Nile, Atbara and main Nile, plus losses to evapotranspiration at Merowe. With its agriculture fully expanded, some authors argue that future water requirements could reach 30–32bcm of water per year (Taha, 2010; Arjoon *et al.*, 2014).

The key question now, and one posed in this chapter, is what strategy will Sudan follow to fulfil its own development interests. Upstream Ethiopia is intent on building generating capacity and a cascade of dams behind the GERD on the Blue Nile, while Egypt is continuing to claim historic rights to its full quota and expects Sudan to fulfil obligations under the 1959 NWA. In the past, Sudan has pursued several strategies including multilateral, through the NBI/ENTRO projects, and unilateral, through expansion of its national infrastructure. Its primary objective has always been maintaining and augmenting storage capacity, most recently the main purpose behind heightening the Roseires Dam. Currently, Sudan is trying to maximize its benefits from the GERD, through trilateral negotiations with the two neighbours. On the table for Sudan is the possibility of coordination between the GERD and Roseires that would allow water releases to be as favourable as possible for the irrigation ambitions of the Sudanese government. However, with this new configuration of opportunities and challenges, Sudan's voice on Nile affairs has begun to be heard more profoundly, particularly as it moves from 'silent positionality' to an active role in the basin. The country is now moving beyond 'mediation' and 'brokering' to asserting its own interest, as the recent statement from Sudan's foreign minister after the GERD trilateral talks suggested in late 2015:

> Sudan is neither a mediator, nor are we neutral or biased, but we have rights just like Egypt and Ethiopia. We are moving with the negotiations in the frame of our national interests that entitle us to converge viewpoints between the three parties.
>
> (*Al-Ahram*, 2015b)

Sudan's two large Nile neighbours will now watch and act with interest, both keen to court Sudan's position on the Nile given that its alliances during negotiations and agreements will be key for directions in the hydropolitical game. Sudan's future ambitions and attitudes can change the old balance of power between upstream and downstream riparians, and can be a determinant in negotiation of new legal agreements and the evolution of cooperation processes in the Nile Basin. New investments can affect the already strained relations between countries – particularly when they are large scale (such as the GERD) and highly water consumptive (such as irrigation in Sudan). Impacts can be even bigger (or unpredictable) knowing they are taking place outside a sanctioned agreement between all parties and not under a multilateral institution, such as the NBI. It is likely that Sudan will continue to develop its consumptive use of water which will progressively impact upon flows to Egypt. Without the 'water loan' from Sudan – in the order of 5bcm annually – major changes will be required in Egypt's irrigation system and this will shrink ambitions for further irrigation expansion. Sudan's increasing water consumption will also 'sensitize' the issue of additional dams upstream constructed in the Ethiopian catchment, which are being planned by Ethiopian authorities. This is likely to push Sudan further into alliances with Ethiopia, keen to gain further access to cheap hydroelectricity while benefitting from more effective river regulation and storage. In short, the evidence suggests an inexorable shift upstream in interests and political gaming, with the kingmaker seeking to influence outcomes. This reflects a wider sea change in Nile Basin politics and economics that has taken place in the first decade of the twenty-first century – a decade of unprecedented changes to the position of states within Nile Basin development.

References

AFP (2014). *Sudan asks Egypt to rejoin Nile Basin body*. 19 June.

Agreement (1959). *Agreement between the Republic of the Sudan and the United Arab Republic for the Full Utilization of the Nile Waters*. Signed at Cairo, Egypt, 8 November.

Agreement (2010). *CFA Agreement on the Nile River Basin Cooperative Framework*. Signed at Entebbe, Uganda, 14 May.

Agreement (2015). *Agreement on Declaration of Principles between the Arab Republic of Egypt, the Federal Democratic Republic of Ethiopia and the Republic of the Sudan on the Grand Ethiopian Renaissance Dam Project*. Signed at Khartoum, Sudan, 23 March.

Al-Ahram (2014). Sudanese FM says his country impartial in Egypt–Ethiopia crisis over Renaissance Dam. 19 February.

Al-Ahram (2015a). Bashir says Ethiopian Renaissance Dam 'has become a reality'. 6 December.

Al-Ahram (2015b). Egypt, Ethiopia, Sudan to hold more Nile dam talks. 12 December.

Arjoon, D., Mohamed, Y., Goor, Q. and Tilmant, A. (2014). Hydro-economic risk assessment in the Eastern Nile River Basin. *Water Resources and Economics*, 8: 16–31.

Awulachew, S. B., McCartney, M., Steenhuis, T. S. and Ahmed, A. A. (2008). *A Review of Hydrology, Sediment and Water Resource Use in the Blue Nile Basin*. International Water Management Institute Working Paper 131. Colombo: IWMI.

Blackmore, D. and Whittington, D. (2008). *Opportunities for Cooperative Water Resources Development on the Eastern Nile: Risks and Rewards. An Independent Report of the Scoping Study Team to the Eastern Nile Council of Ministers*. Washington, DC: World Bank.

Block, P., Strzepek, K. and Rajagopalan, B. (2007). *Integrated Management of the Blue Nile Basin in Ethiopia: Hydropower and Irrigation Modeling*. IFPRI Discussion Paper 700. Washington, DC: International Food Policy Research Institute.

Cascão, A. E. (2008). Ethiopia: Challenges to Egyptian hegemony in the Nile basin. *Water Policy*, 10(2): 13–28.

Cascão, A. E. (2009). Changing power relations in the Nile River Basin: Unilateralism vs. cooperation? *Water Alternatives*, 2(2): 245–268.

Cascão, A. E. (2012). Nile water governance. In Awulachew, S. B., Smakhtin, V., Molden, D. and Peden, D. (eds). *The Nile River Basin: Water, Agriculture, Governance and Livelihoods*. New York: Routledge, 229–252.

Collins, R. O. (2008). *A History of Modern Sudan*. Cambridge: Cambridge University Press.

Conniff, K., Molden, D., Peden, D. and Awulachew, S. B. (2012). *Nile Water and Agriculture: Past, Present and Future.* In Awulachew, S. B., Smakhtin, V., Molden, D. and Peden, D. (eds). *The Nile River Basin: Water, Agriculture, Governance and Livelihoods.* New York: Routledge, 5–29.
DIU (Dams Implementation Unit) (2008). *President Al-Bashir: Heightening of Roseires Dam is Sudan's Real Oil,* 16 November. Available online at www.roseiresdam.gov.sd/en/index.php/home_en/show/60.
Eissa, S. (2008). International law and hydro-hegemony in the Nile Basin: A Sudanese perspective. *Water Policy,* 10(2): 29–49.
Eldaw, A. and Fekade, W. (2009) *Sustainable Transboundary Basin Development as a Strategy for Climate Change-Induced Conflict Prevention: Reflections from Eastern Nile.* Paper presented at the Workshop on Climate Change and Transboundary Water Resource Conflicts in Africa, 29–30 September, Mombasa, Kenya.
El-Zain, M. (2007). *Environmental Scarcity, Hydropolitics, and the Nile.* Maastricht: Shaker Publishing.
ENSAP (2008). *ENSAP Projects Brief Status Report and Future Activities, April 2008.* Addis Ababa: ENTRO Office.
ENSAP (2009). *Factsheets of the 8 IDEN Projects.* Addis Ababa: ENTRO Office.
Ethiopian News (2011). Ethiopia launched grand millennium dam project, the biggest in Africa. 2 April.
Faki, H. H. M., Gumaa, Y. T. and Ismail, M. A. (1995). Potential of the Sudan's irrigated sector in cereal grains production: Analysis of various policy options. *Agricultural Systems,* 48: 457–483.
FAO (2011). *International Investments in Agriculture in the Near East.* Cairo: FAO Regional Office for the Near East.
FAO AQUASTAT (2015). *Sudan's Profile.* Available online at www.fao.org/nr/water/aquastat/countries_regions/sdn/index.stm.
Geressu, R. T. and Harou, J. (2015). Screening reservoir systems by considering the efficient trade-offs: Informing infrastructure investment decisions on the Blue Nile. *Environmental Research Letters,* 10(12): 125008.
Guariso, G. and Whittington, D. (1987). Implications of Ethiopian water development for Egypt and Sudan. *International Journal of Water Resource Development,* 3(2): 105–114.
Hamad, O. E. and El-Battahani, A. (2005). Sudan and the Nile basin. *Aquatic Sciences,* 67(1): 28–41.
IPoE (2013). *Final Report on the Grand Ethiopian Renaissance Dam,* 31 May, Addis Ababa, Ethiopia.
Johnston, R. (2012). Availability of water for agriculture in the Nile Basin. In Awulachew, S. B., Smakhtin, V., Molden, D. and Peden, D. (eds). *The Nile River Basin: Water, Agriculture, Governance and Livelihoods.* New York: Routledge, 61–83.
Kahsay, T. N., Kuik, O., Brouwer, R. and Van der Zaag, P. (2015). Estimations of the transboundary economic impacts of the Grand Ethiopian Renaissance Dam: A computable general equilibrium analysis. *Water Resources and Economics,* 10: 14–30.
Keulertz, M. (2012). *Drivers and Actors in Large-Scale Farmland Acquisitions in Sudan.* Land Deals Politics Initiative. PLAAS: Cape Town.
Knott, F. and Hewitt, R. (1994). Future water development planning in Sudan. In Howell, P. and Allan, J. A. (eds). *The Nile – Sharing Scarce Resources: A Historical and Technical Review of Water Management and of Economic and Legal Issues.* New York: Cambridge University Press, 205–216.

Liu, H., Masera, D. and Esser, L. (eds) (2013). *World Small Hydropower Development Report 2013*. UNIDO: International Center on Small Hydro Power.

MAF (Ministry of Agriculture and Forests, Sudan) (2000). Existing and future irrigated area along the Nile tributaries and water consumption. In O. E. Balgis. (ed.) *Data and Information on Environment and related issues in Eastern Nile Sub-Basin in Sudan*. Report prepared for ENTRO, April 2006.

Mahgoub, F. (2014). *Current Status of Agriculture and Future Challenges in Sudan*. Current African Issues, 57. Uppsala: Nordiska Afrikainstitutet.

McCartney, M., Alemayehu, T., Easton, Z. and Awulachew, S. (2012). Simulating current and future water resources development in the Blue Nile River Basin. In Awulachew, S. B., Smakhtin, V., Molden, D. and Peden, D. (eds). *The Nile River Basin: Water, Agriculture, Governance and Livelihoods*. New York: Routledge, 269–291.

MIT (Massachusetts Institute of Technology) Non-partisan Eastern Nile Working Group (2014). *The Grand Ethiopian Renaissance Dam: An Opportunity for Collaboration and Shared Benefits in the Eastern Nile Basin – An Amicus Brief to the Riparian Nations of Ethiopia, Sudan and Egypt*. Convened by MIT Jameel World Water and Food Security Lab, 13–14 November.

Mohieldeen, Y. (2007). *Sudan's Nile Waters and the Eastern Nile Basin: Hydropolitics in a Highly Politicized Environment*. PhD thesis. London: School of Oriental and African Studies, University of London.

Nazir, O. A. and Desai, G. D. (2001). *Kenana: Kingdom of Green Gold – Grand Multinational Venture in the Desert of Sudan*. London: Kegan Paul.

NBI (2012). *Sudan, the NBI and the Benefits of Cooperation*. NBI Country Papers. Entebbe, Uganda: NBI.

NBI (2014). *Nile Cooperation: Opportunities and Challenges*. NBI Flagship Paper. Entebbe, Uganda: NBI.

Nicol, A. and Cascão, A. E. (2011). Against the flow: New power dynamics and upstream mobilisation in the Nile Basin. *Review of African Political Economy*, 38(128): 317–325.

Reuters (2013). Sudan launches major dam to boost agricultural production investment. 1 January.

RoS (Republic of Sudan) (2008). *Action Plan for Agricultural Revival, High Committee for Studying the Present Status of Agriculture and Articulate a Future Vision and Executive Programme for Agricultural Revival*. Khartoum: Republic of Sudan.

RoS (n.d.). *Investment in Sudan Agriculture's Sectors, in webpage of Sudan Embassy in the UK*. Available online at www.sudanembassy.co.uk/index.php/about-us/about-sudan/business-economy/580-agriculture-s-sectors.

Salman, S. M. (2011). The new state of South Sudan and the hydro-politics of the Nile Basin. *Water International*, 36(2): 154–166.

Satti, S., Zaitchik, B. and Siddiqui, S. (2015). The question of Sudan: A hydro-economic optimization model for the Sudanese Blue Nile. *Hydrology and Earth Systems Sciences*, 19: 2275–2293.

SD (*Sudan Vision*). (2012a). The white blessing back on track: Cotton returns as strategic crop. 1 May.

SD (2012b). Roseires Dam doubles in height. 6 November.

SD (2013a). Investment can take place in already existing projects to increase production, Interview. 20 May.

SD (2013b). Turkey to invest $300mln in agriculture in Sudan. 5 July.

SD (2013c). Establishment of number of free zones at states bordering South Sudan. 14 August.

SD (2013d). Agriculture is the sole way to boost national economy. 7 December.
SD (2014). Farmers' union announces support to president food security initiative. 19 January.
ST (*Sudan Tribune*). (2008). Sudan, China sign $396mln contract to heighten Roseires Dam. 28 April.
ST (2010). Sudan freezing its membership in the Nile basin initiative, 27 June.
ST (2011a). Ethiopia: Govt blames ousted Egyptian regime over Nile water crises. 30 April.
ST (2011b). Sudan: Country agrees to tripartite committee over Ethiopia's Nile Dam. 25 October.
ST (2012a). Sudan: Minister, Egypt and Ethiopia to meet for talks over Blue Nile Dam. 26 February.
ST (2012b). Sudan's Bashir supports Ethiopia's Nile Dam project. 8 March.
ST (2012c). Agreement reached on Eastern Nile Basin cooperation. 9 November.
ST (2013a). Sudan downplays negative impact of Ethiopian Dam project. 30 May.
ST (2013b). Egypt TV airs live discussion of Ethiopia Dam sabotage plans. 4 June.
ST (2013c). Sudan reiterates support of Ethiopian Dam plans. 9 June.
ST (2014a). Sudan FM criticises Egypt over Ethiopian Dam dispute. 18 February.
ST (2014b). Egypt wants Sudan to mediate in Nile water tripartite meeting. 24 August.
ST (2014c). Ethiopia's Nile Dam project nears halfway mark. 10 December.
ST (2015a). Egypt accuses Ethiopia of stalling talks on Nile Dam project. 8 December.
ST (2015b). Sudan, Egypt and Ethiopia reach agreement on Renaissance Dam. 29 December.
Stads, G. and El-Siddig, K. (2010). *Sudan: Recent Developments in Agricultural Research*. ASTI Country Notes. Washington, DC: International Food Policy Research Institute.
Taha, F. (2010). The history of the Nile waters in the Sudan. In Tvedt, T. (ed.) *The River Nile in the Post-Colonial Age: Conflict and Cooperation among the Nile Basin Countries*. London: I.B. Tauris, 179–216.
Tvedt, T. (2004). *The River Nile in the Age of the British: Political Ecology and the Quest for Economic Power*. London: I.B. Tauris.
Tvedt, T. (2013). *A Journey in the Future of Water*. London: I.B. Tauris.
Verhoeven, H. (2012). Sudan and its agricultural revival: A regional breadbasket at last or another mirage in the desert? In Allan, T., Keulertz, M., Sojamo, S. and Warner, J. (eds). *Handbook of Land and Water Grabs in Africa: Foreign Direct Investment and Food and Water Security*. London: Routledge, 43–56.
Warner, J. (2013). The Toshka mirage in the Egyptian desert: River diversion as political diversion. *Environmental Science and Policy*, 30: 102–112.
Waterbury, J. (1979). *Hydropolitics of the Nile Valley*. Syracuse: New York University Press.
Waterbury, J. (2002). *The Nile Basin National Determinants of Collective Action*. New Haven, CT: Yale University Press.
Whittington, D., Waterbury, J. and Jeuland, M. (2014). The Grand Renaissance Dam and prospects for cooperation on the Eastern Nile. *Water Policy*, 16(4): 595–608.

7 Transboundary water resources and the political economy of large-scale land investments in the Nile

Sudan, hydropolitics and Arab food security

Ramy Lotfy Hanna

Introduction

The end of the status quo in the Nile Basin

Egypt and Sudan are two key downstream countries in the Nile transboundary river basin. A significant part of the Nile Basin's ecosystem and its watershed's natural resources within both countries were under the control of the Ottoman Empire and its succeeding khedives and kings since Mohamed Ali's invasion of Sudan in 1821. With the increased British influence in Egypt and Sudan in 1882 and 1897, respectively, coupled with colonial strategic interests in both countries – especially cotton cultivations and trade (Waterbury, 1979) – the Nile River flow was governed by the 1929 Agreement.[1,2] Sudan eventually gained its independence in 1956 and signed the 1959 Nile Treaty[3] between Colonel Abboud and Nasser, whereby Egypt and Sudan were granted full rights for the utilization of the Nile waters with annual shares of 55.5 and 18.5bcm (billion cubic metres), respectively.

Today, the Nile River Basin is far from its hydropolitical shape of the nineteenth and twentieth centuries during the eras of the Ottomans, British colonialism and the Anglo-Egyptian Sudan (1899–1956). Sudan[4] is now split into two countries: Sudan and South Sudan.[5] Sudan's quota based on the 1959 Nile Agreement also includes South Sudan's share, with no clear rules on how this allocation will be divided between both countries[6] – a fundamental question that has not been addressed yet in the complex hydropolitics of the Nile River Basin. Egypt and Ethiopia are also key countries likely to affect and be affected by transboundary water resources politics related to both Sudan and South Sudan. Egypt's population grew by 46 per cent between 1994 and 2014 (Youssef *et al.*, 2014), currently at 87 million citizens (CAPMAS, 2014a), forecasted to exceed 100 million by 2025, mostly concentrated within the narrow Nile Valley constituting 6 per cent of the country's total area dominated by the desert. While the available per capita water resources have reached an annual rate of 663 cubic metres/inhabitant in 2013, forecasted to reach 582 cubic metres in 2025 (CAPMAS, 2014b), Egypt's annual share from the

Nile based on the 1959 Agreement has been facing political contestations by upstream countries led by Ethiopia – the source of the Blue Nile contributing 84 per cent of Egypt's share of water resources. Ethiopia is also witnessing a demographic boom forecasted to reach 124 million by 2025 (UNDESA, 2012) with serious socio-economic and poverty challenges. Both Sudan and Ethiopia are undertaking several mega infrastructure developments accompanied by increased unilateral investments in dams, water resources projects and large-scale irrigation (LSI) schemes with the active participation of the private sector. The Nile River Basin is indeed witnessing 'the end of its status quo' (Verhoeven, 2013).

Perhaps the tripartite agreement signed between Egypt, Ethiopia and Sudan in Khartoum on 24 March 2015 indicates a new chapter of cooperation in the Nile. Yet, it is just a beginning that requires a myriad of institutional arrangements and political will to address the various social, economic and environmental dynamics and complexities of the basin and its natural resources. For instance, since the signing of the initial agreement, a series of meetings took place between the Tripartite Committee of Experts of the Grand Ethiopian Renaissance Dam (GERD) consisting of 12 members from the three Blue Nile countries. Up until July 2015, seven meetings have been held with no clear agreement on setting a date to sign a contract with the French and Dutch consulting companies in charge of conducting detailed studies about the hydrological and environmental impacts of the dam (as widely published by several official sources in local newspapers and media outlets). Even during the seventh meeting held on 22 July in Khartoum, negotiations have been reported to face serious bottlenecks whereby the committee has also failed to reach the basic standards that will be put forward to the consulting companies in charge of conducting the assessments, expected to last for 11 months.

But even if these issues are resolved, a new dimension is arising in the Eastern (Blue) Nile Basin concerning longstanding and complex hydropolitics beyond the upstream–downstream contestations of historical water shares and the *state-centric* conflict and cooperation narratives in the transboundary water resources literature. Large-scale land acquisitions (LSLAs) and water-related investments (financing in infrastructure, dams, reservoirs, canals, irrigation and agriculture technologies) are witnessing a different pattern in Africa and the Nile Basin, shaped by principles of the neo-liberal global economy and different political economy drivers. Financial investments in agriculture targeting LSLAs are primarily driven by available fertile land and water resources, especially in countries with similar resource endowments such as Sudan. Such practice can be best described as 'the large scale acquisition of land in developing economically and politically weak countries through FDI by powerful, developed, economically robust, but water-stressed, nations' (Sebastian and Warner, 2013: 2). Overall, 'there is a perceived general pattern that investors do not seek lands that do not have water for production in the first place. So, land in itself is meaningless for their purpose without water' (Mehta *et al.*, 2012: 196). Despite sharp disagreements about the scholarly approaches to define, calculate and report large-scale land deals (see Scoones *et al.*, 2013), trends and patterns clearly highlight that African governments supply a large share of these acquisitions or 'grabs'. Available evidence indicates that much of the announced land deals did not actually materialize; nevertheless, they point to an increasing trend for the use of land and water resources abroad.

In light of the increased government support and promotion of agricultural investments, evidence suggests that many African countries are 'motivated by potential revenues from water fees and the prospect of improved agricultural productivity, signing away water rights for decades to large investors' (IIED, 2011: 1). Despite the fact that implementation on the ground is significantly less than the announced deals (Verhoeven, 2013; Woertz, 2013a), the documented trend of LSLAs by the different state and non-state actors is captured by the term 'water grabs' (Mehta *et al.*, 2013). That is,

> while control over water resources has been traditionally associated with state control and domination by national rulers (cf. Wittfogel 1957; Worster 1983) the term water grabbing draws attention to the involvement of new capitalist players and actors in water resources management and the rise of new political and economic power relations through diverse trajectories of neo-liberalism.
> (Mehta *et al.*, 2013: 198)

This chapter highlights the increasing role of state and non-state actors in acquiring land and 'water resources' in the Nile transboundary river basin.

Water grabs as a concept has been typically addressed from the lenses of social justice, livelihoods and equity on local and national scales, but rarely, if ever from a 'hydropolitical' standpoint. Hydropolitics here is understood as 'the systematic investigation of the interaction between states, non-state actors and a host of other participants, like individuals within and outside the state, regarding the authoritative allocation and/or use of international and national water resources' (Meissner, 1999, in Turton, 2000). In this respect, it is important to understand the connotations of water grabs on the larger scale of transboundary river basins 'over water resources that transcend international borders' (Elhance, 1999: 3). This chapter seeks to do so with a special focus on the case of Gulf Cooperation Council (GCC) and Arab agricultural investments in Sudan motivated by official food and water security policies.

Given this background, the aim of the chapter is to contribute to the current debates on transboundary water resources *beyond the state-centric discourse*, notably as it relates to the role of *non-state actors* and *state-capital alliances* in LSLAs and hydropolitics in the Nile Basin. The chapter examines the political economy drivers and the different actors involved in public and private finance investments, notably from Gulf (GCC) countries and China among others (i.e. Brazil, India, South Korea and Jordan) targeting LSLAs in transboundary river basins, with a special focus on the case of Sudan in the Nile Basin. In this respect, the analysis addresses the hydropolitical relations between Egypt and Sudan, on the one hand, and the increasing Chinese, GCC and Arab agricultural investments in Sudan, on the other, as part of the larger policy objectives seeking to implement the Arab water security and the Arab food security strategies.

Analytical framework and methodology: a constructivist approach

The analytical framework aims therefore to examine the agency of non-state actors in global water security, management and governance (Sojamo and Larson, 2012) on the geographic scale of transboundary river basins. Agency here is seen as 'the capacity of individual and collective actors to change the course of events or the outcome of processes' (Pattberg and Stripple, 2008 in Sojamo and Larson, 2012: 620). Specifically, the analysis builds on Ken Conca's (2006) framework for understanding global environmental governance based on key institutional presumptions about 'territoriality, authority, and knowledge' suggesting that 'international regimes are embedded in a nation-state system poorly suited to an engagement with global environmental challenges that are territorially unconstrained, incompletely illuminated by natural sciences, and intensely political' (Gareau and Crow, 2006: 318). Such proposition is perhaps relevant to today's global economy where the increasing interdependencies between food security, (virtual water) trade, investments and capital could have deep implications for natural resources elsewhere. The Nile Basin is a good example where different exogenous players are taking part in the use and utilization of natural resources, notably land and water resources.

Sudan is an interesting case study for several reasons. First, Sudan is metaphorically and physically in the middle of the hydropolitical complex relations in the Eastern (Blue) Nile River Basin. Second, Sudan gains significant importance among the other Nile Basin countries due to its size and the area of the river basin within its geographical territory. All parts of Sudan including both Sudan and South Sudan represent a total area of 2,551,341km^2, where *Sudan*'s landmass is 1,911,341km^2 compared to 640,000km^2 in South Sudan (Nile Basin Initiative, 2012). In terms of water resources, it is interesting to realize that 2,062,558km^2, equivalent to 80.84 per cent of both countries (all parts of Sudan), are within the basin, where more than 64 per cent of the basin's total area (3,176, 543km^2) is within both countries (Nile Basin Initiative, 2012).

Sudan is also hydropolitically bound to the 11 Nile Basin countries and is also politically committed and plays an active role in the Nile Basin Initiative (NBI), with a recent project of hydropower for joint electricity generation with Ethiopia. In addition, Sudan is viewed as the 'breadbasket' of the Arab World: a vision that witnessed unsuccessful past experiences in agricultural development and contemporary attempts of revival. In this sense, Sudan can be viewed as wearing two hats: a key country in Africa's Nile River Basin, and a cornerstone of the Arab food security strategy emphasized through its participation and role in the League of Arab States (LAS). Particularly in Sudan, a number of domestic and foreign investors are increasingly invited to invest and acquire massive land areas, and, consequently, water resources, including underground and surface waters, rain-fed agriculture and green water. The chapter provides evidence about existing and planned large-scale agricultural investments in Sudan by a multitude of stakeholders representing state and non-state actors highlighting the role of financial investments in appropriating natural resources, with special attention to China's role in infrastructure development and GCC agricultural investments.

The research realizes that despite the efforts of various media and academic reports to spot and document land acquisitions and grabs in different parts of the world, little is still known about the details of these deals and projects, where accuracy is often questioned. This is perhaps best manifested in the discrepancy between the amounts of announced land deals vis-à-vis the actual level of implementation on the ground, often fitting one of these three categories: projects that did not start, projects in early stages of development or projects where implementation took place only on a fraction of the intended original size (Woertz, 2013a). Accordingly, presented data has been collected from primary and secondary resources, including official government documents in Arabic and constructed databases. A stakeholder mapping has also been undertaken to identify different state and non-state actors, their institutional affiliations as well as their ongoing and planned agricultural investments in Sudan. Interviews were also held with a number of Sudanese, Egyptian and international experts from the Food and Agriculture Organization (FAO), the Common Market for Eastern and Southern Africa (COMESA) and the NBI, among others. The research process has also benefited from the interaction with the Land Matrix (www.landmatrix.com) and its committed staff in charge of capturing land deals and corresponding water resources in different parts of Africa and Sudan.

The rest of the chapter is structured as follows: the second section provides an overview of the agriculture sector and existing LSI schemes in Sudan. The chapter explores the ongoing and planned water resources infrastructure development and agricultural investments relying on Chinese expertise and Arab financing. The third section discusses the intersection of Sudan's breadbasket vision, on the one hand, and the Arab food and water security strategies, on the other. The section also provides a detailed analysis of the nature and determinants of recent land demand by 'state and non-state actors' where different 'investment modalities' are presented with examples of domestic and foreign water-land investments. The section also highlights the different challenges facing agricultural investors in Sudan. The fourth section analyses LSLAs from a transboundary water resources lens and provides an analysis of the delicate hydropolitics in the Nile Basin given the growing 'state-capital alliances' and the increasing role of state and non-state actors in acquiring natural resources via financial capital investments. This section also conceptualizes the governance implications of LSLAs in transboundary water resources and how the governance in the Nile as a transboundary basin could be changing given the increased state and non-state actors' participation in land and water investments. The final section concludes the chapter.

Water resources, infrastructure and agricultural investments in Sudan

Existing large-scale agricultural schemes

Historically, unleashing Sudan's full agricultural potential has been a challenge. Most agricultural activity has been concentrated in three areas: Khartoum, the central provinces and the northern region, attracting 86 per cent of foreign direct investment (FDI) partly due to available infrastructure, while other regions east and west of the country have been neglected (Geopoliticalmonitor, 2013). Sudan has roughly 650,000km^2 of land with high agricultural potential, but 75 per cent of it is farmed at only 10–50 per cent of its capacity, and one-quarter of the land that has high suitability is farmed at only 10 per cent of its capacity (Ranganathan and Briceño-Garmendia, 2011). According to the FAO, the country's agricultural policy was changed in 1997 in an effort to attain greater food self-sufficiency by reducing the area of cotton production due to irrigation water shortage, replaced instead by wheat and sorghum, which require less water (Mahgoub, 2014: 13). Overall, five main types of agricultural activities exist in Sudan: mechanized rain-fed, traditional rain-fed, mechanized irrigation, traditional irrigation and livestock husbandry/pastoralism (UNEP, 2007). Irrigated area and existing LSI include several federal schemes such as Gezira, Rahad, New Halfa, Souki, Guneid and Kenana (Embassy Press, 2015), among others. 'Al-Gezira' irrigation scheme – located south of Khartoum between the Blue and White Nile – is considered one of the largest in the world under a single administration covering nearly half of Sudan's total irrigated area and consumes 35 per cent of its Nile water share (UNEP, 2007). Cotton cultivation was prioritized for nearly 80 years[7] with disappointing cropping intensities and poor yields partly due to government policies, poor and centralized canal maintenance, lack of irrigation water, inefficient water use, and little incentive for farmers to improve production because cotton as a mandatory crop is financed and marketed by the government (Molden *et al.*, 2009). As a result, about 12 cotton gins closed due to lack of financial support and neglect (Dabanga, 2014). During a press conference on 30 November 2014, President Al Bashir had described the scheme as 'non-feasible, and a burden on the country's budget since the end of the 1960s' indicating that 'farmers always want the government to fund their agricultural projects without paying back their loans' (*Sudan Tribune*, 2015). The other two major schemes are the Rahad on the bank opposite Al-Gezira, intended to use water from the Roseires Dam, and New Halfa on the Atbara River using water from the Khashm el Girba Dam. Both schemes are often described as poorly planned and are plagued by siltation and irrigation inefficiency with below average yields (UNEP, 2007; Awulachew *et al.*, 2012). In addition, there are five major sugar schemes, four of which are government run. The fifth and largest sugar plantation is the Kenana Sugar Company (KSC), which is an international public–private joint venture (UNEP, 2007). Overall, Sudan illustrates the case of a fairly land-abundant country that promoted large-scale agricultural investment in response to the 1970s oil shock with mixed results (Deininger and Byerlee, 2011).

Arab financing and China's role in water resources infrastructure development

Despite the availability of land areas solely dependent on the abundance of rain for agriculture, Sudan has embarked on an ambitious plan to improve its irrigation system. Dams are being built throughout the country diverting a large share of water resources towards new irrigation projects. Aiming to attract local and foreign capital, the government is prioritizing agriculture for self-sufficiency through new infrastructure investments to facilitate easy access to irrigation from the Nile. Khartoum has staked its political future in the post-oil era on the most far-reaching (hydro) infrastructure programme in Sudanese history, with Kuwaiti, Emirati and Qatari funding for its dams totalling almost US$10 billion (Verhoeven, 2011: 695–699; Verhoeven, 2013: 12). According to the geopolitical monitor, early in 2013 Gulf nations contracted a Chinese firm to expand the Roseires Dam, substantially increasing Sudan's ability to irrigate farmland. This is perhaps similar to earlier investments during the 2000s when the Abu Dhabi Fund for Development (ADFD) lent Sudan a total of US$225 million to support the Merowe and Roseires dam projects (EIU, 2015). Indeed, water resources infrastructure developments such as dams for hydropower and irrigation are increasingly taking place in Sudan, as shown in Figure 7.1, benefiting from international expertise, foreign financing and private-sector participation. In 2003, the government initiated the US$1.2 billion Merowe Dam hydropower project inaugurated in April 2009 with Chinese support (International Rivers, 2011a; Verhoeven, 2014: 63). The dam's total capacity is 12.5km^3, around 20 per cent of the Nile's annual flow, producing 1,250 megawatts (MW) (EJATLAS, 2014a). Different international financial institutions participated in financing the project, including China Export Import Bank, Kuwait Fund for Arab Economic Development, Saudi Fund for Development, Arab Fund for Economic and Social Development (AFESD) from Kuwait, United Nations Missions in Sudan, Fund for Development of Youth Projects or Sharakah, and the ADFD from the United Arab Emirates (UAE) (EJATLAS, 2014a). Different companies and state enterprises also participated in implementing the project, such as the German Lahmeyer International, Alstom from France, ABB Power, HPE, and Jilin Province Transmission Substation Project Company from China. The government is also promoting the Merowe agricultural project with a main objective to realize food security for Sudan and the Arab World. The project covers the area along the Nile banks, starting at the fourth waterfall through Semit Island at the vicinity of the third waterfall (north) including upper terraces of the Nile River west Dongola and Alkaab Valley (COMESA Regional Investment Agency (RIA, 2013). The main source of water is the Nile from the Merowe reservoir where the left bank areas are irrigated by a 415km canal called 'Left Bank Canal'; the right bank areas are irrigated by a 157km canal called 'Right Bank Canal', in addition to five plants as Nile water intakes. The total estimated area of the project is 904,000 feddans (380,000ha) divided as follows: area of the Right Bank Canal (26,551.8ha); area of the Left

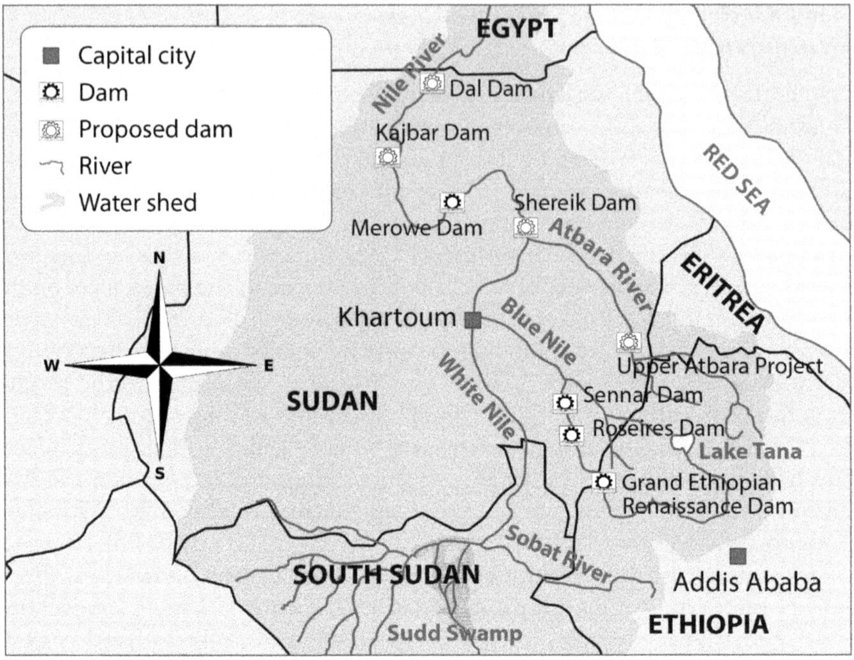

Figure 7.1 Map displaying major dam projects in Sudan

Source: Henrik Alfredsson.

Bank Canal (230,606.4ha); pump irrigation areas (121,230.4ha). According to the Ministry of Investment, work had started under the 'early execution plan' in West Kould, an area of 200,000 feddans with a value of US$500 million using Chinese credit (National Investment Authority, 2013).

Other new dam projects include the Dal-Kajbar dams, two separate dams located at the second and third cataracts, respectively. Kajbar Dam is located north of Merowe and Dongola City. The construction of Kajbar, a concrete dam of 2.6km and five generating units with a total capacity of 320 megawatts (MW), was awarded in 2010 to the Chinese company Sinohydro, the world's largest hydropower contractor. The project is controversial as it floods 90 villages, displaces 10,000 people and destroys 500 archaeological sites, leading to peaceful protests in 2007 due to a lack of consultation of the local population (International Rivers, 2011b). The project is implemented by Sudan's State Enterprise Kajbar Electricity Co.,[8] in participation with Chinese Harbin Power Engineering Company and Sinohydro Corporation Limited (EJATLAS, 2014b). The Dal Dam is the northernmost of the dams in Sudan planned to be located less than 40km northeast of Sai. According to the Anti Dal-Kajbar Committee, there are two scenarios for the dam, Low Dal and High Dal, where in the former all points below 201 metres above sea level (masl) will be submerged, and for the second option an artificial lake is expected to submerge everything below 219msal. The dam has been highly controversial, leading to social mobilization and protests (Committee of Anti Dal-Kajbar Dams, 2011).

The Shareek Dam Project in the River Nile province (3.5km concrete dam, 500km north of Khartoum city near the fifth cataract) aims to generate 420MW of hydroelectric power for local demand, while transmitting the surplus to the national grid, in addition to its irrigation potential. Furthermore, the Upper Atbara and Setit Dam Complex under construction since 2011 comprises Rumela Dam on the Upper Atbarah River and Burdana Dam on the Setit River in Eastern Sudan[9] (Awulachew *et al.*, 2012), expected to be completed by spring 2016 (Preservenile, 2012). The project's objective is to support agricultural development, hydropower generation and access to potable water, with a total estimated cost of US$1.9 billion, out of which US$838 million is allocated for dams construction by the China Three Gorges Corporation and its overseas project-contracting subsidiary China Water and Electric Corporation. Further financing for the dam was sought through a concessional loan of US$90 million by the ADFD, as announced in February 2015 during Sudan's presidential visit to the UAE (EIU, 2015).

Overall, Sudan has heavily invested in infrastructure in recent years. With the Merowe Dam providing an additional capacity of 1,250MW since 2009, power generation capacity is forecasted to increase due to additional hydropower investments (Ranganathan and Briceño-Garmendia, 2011: 2); yet service reliability remains an issue. Two-thirds of Sudan's spending consists of infrastructure investment reaching approximately US$1.5 billion annually, about 6 per cent of its gross domestic product (GDP). Chinese funds account for 40 per cent of investment finance in the country, private investments in infrastructure represent 23 per cent of the total, while official development assistance (ODA) is negligible (Ranganathan and Briceño-Garmendia, 2011: 2). Nevertheless, Sudan's annual infrastructure funding gap remains US$2.9 billion, equivalent to 14 per cent of its GDP. To bridge this gap, Sudan is seeking to mobilize financial resources from Arab countries, with technical expertise from China and the private sector.

Supply side of land and water investments in Sudan

In addition to the existing agricultural schemes and planned dam infrastructure projects, Sudan is seeking to promote itself as an agricultural investment destination targeting local and foreign capital. From the supply side, Sudan is offering millions of feddans of fertile land depending on available water resources from the Nile River, underground aquifers, rain-fed agriculture and green water. The process of land allocation is described as follows:

> Federal and state ministries of agriculture and investment commissions can allocate large concessions. Allocation of land is the main mandate of the Government Agricultural Land Disposition Committee (GALDC), which applies the procedures for land allocation. To obtain agricultural leasehold, an applicant submits an application to the state's governor explaining the

> intentions behind and purpose for requesting the land, provides proof of financial capacity, and indicates the location and size of the requested area, among other information. The governor then transfers the application to the GALDC, and the Ministry of Agriculture makes a technical inspection on-site to check whether the proposed project is viable. After the GALDC approves the application, and in the absence of contestation during the publicity period, dues are paid to the Land Department and the land lease contract can be signed and sent to the Land Registration Office for recording.
>
> (Deininger and Byerlee, 2011)

In 2013, the Sudanese National Investment Authority and the Presidency Office issued the *Directory of Proposed Investment Projects*, including 323 projects in different economic sectors with a total investment cost of US$30 billion. Out of these 323 projects, 117 projects are in the agriculture sector, totalling 8 million acres distributed among 12 states. Table 7.1 shows the number of projects offered by different states and the total land area in acres. The

Table 7.1 Number of projects offered by different regional states in Sudan and their comprised land area

State	*Number of projects in agricultural sector*	*Total area in acres**
Khartoum State	12	278,624
Northern State	8	1,238,000
River Nile State	11	3,272,500
Kasala State	3	75,000
Gadaref State	–	–
Red Sea State	3	NA
Gezira State	9	959,100
White Nile State	11	1,250,000
Sennar State	15	1,114,000
Northern Kordofan State	5	210,000
North Darfur State	2	NA
South Darfur State	3	NA
Central Darfur State	3	NA
Total	85	8,397,224 acres equivalent to 3,398,236 hectares

Note
* Author calculations based on National Investment Authority data.

Source: National Investment Authority (2013).

directory also lists four national institutions planning 27 new investments in the agriculture and agro-industry sectors: Construction Fund for Eastern Sudan (20 projects); Sundus Agricultural Project (four projects); KSC (two projects); and Africa City of Technology (one project). Under this investment plan, government land is advertised 'with no restrictions and to be registered with the services of water, electricity, and roads available' (National Investment Authority, 2013: 4). Interestingly enough, the document divides agricultural investments into three categories:

1 agricultural investments attracting foreign investors;
2 agricultural investments targeting public-sector projects;
3 agricultural investments led by the private sector but in partnership and participation with farmers, shepherds and fishermen.

Agricultural investments projects are broadly defined under the following categories: sugar projects; agricultural crops projects; rice projects; animal feed projects; fruits and legumes projects; and sunflower projects. For example, in the Northern state the Integrated Agricultural Mechanization Project aims to develop the technological and agricultural mechanical capacities through direct investment. The West El-Goled Project is a proposed undertaking on 50,000 acres on the west bank of the Nile south of Dongola City with proximity to a major airport. The El-Khoy Project on the eastern bank of the Nile – on the opposite bank of El-Goled Project – is planned on 200,000 acres adjacent to the Nile with easy access to water for irrigation. The Wadi El-Khoy Project planned on 250,000 acres focuses on animal husbandry for international export and domestic markets, also taking advantage of proximity to the Nile for its supply of water.

Other large-scale agricultural projects are also planned in the southern and central regions. In the White Nile State, south of the capital, the El-Seela Project is in its early stages of implementation on an area of 150,000 acres targeting a variety of crops, including the highly exportable Galia melon. The project's central location allows it to benefit from available infrastructure and major national roads. In Sennar State, east of the White Nile, additional opportunities for agricultural investment include the 75,000 acre Blue Nile Sugar Project aiming to cultivate and produce sugar cane. Other projects in Sennar include the El Rammash Project, currently producing cotton, sunflower and sorghum grasses; however, as in the Blue Nile Sugar Project, the government seeks a transition to sugar cultivation and requires investors' support to provide farming inputs. Also in Sennar is the proposed Cuziera Project planned on 60,000 acres and targeting rain-fed crops[10] where the government is planning to integrate local communities and farmers by providing employment and land-sharing agreements. In the capital region of Khartoum, plans include establishing the El-Ferdous Project on 260,000 acres, to be irrigated from either the Nile or the Nubian Basin underground water resources. In Al-Gezira State, relatively modest investment opportunities include the 20,000 acre Sharq-El

Nil Project for feed products and red meat to supply local and regional markets. On the southern border of Gezira, north of Sennar, the West El-Hosh Project is proposed on 32,000 acres. Other projects include the Upper Atbara Agricultural Project located east of the Upper Atbara River and west of Kasala City planned on 500,000 feddans (208,333ha) depending on water resources from the Upper Atbara and Stet Rivers.

Sudanese natural resources and Arab food and water security: the role of state and non-state actors

Available 'water resources' are a key driver in attracting foreign agricultural investments in Sudan. Fundamentally, it is important to understand that Egypt and Sudan are two key countries in the Nile Basin, but also share a prominent status in the LAS with a common vision of achieving joint Arab economic integration in trade and food security (Abdel Wahab, 2011), along with water-scarce Gulf (GCC) countries. This vision was emphasized during President Bashir's speech at the opening ceremony of Egypt's Economic Development Conference in Sharm El Sheikh on 13 March 2015, where he invited Arab financing to invest in Sudan's available land and water resources, stating that 'Sudan has the land and water, Egypt has the human resources and expertise, while gulf countries can provide financing' (Al Bashir, 2015). Despite the past disappointments in Sudan's 'breadbasket' vision initially perceived by Numeiry in the 1970s (Verhoeven, 2015), the dream of Sudan's agricultural revival is still alluring.

As one of the very few Arab economies where horizontal agricultural expansion is possible due to the availability of land and water resources, Sudan is typically labelled as 'the breadbasket of the Arab World'. Agriculture production is the cornerstone of the economy with a wealth of fertile land representing 47.4 per cent of the total area in 2012 (World Bank, 2015) and employs around 60–80 per cent of the workforce, contributing nearly 34.5 per cent of the GDP in 2013 (ADB *et al.*, 2015). Yet, successive governments were not able to mobilize financial resources to make use of the country's comparative advantages, leading to a clear contradiction as Sudan relies on imports for its food security. This is often attributed to several factors:

- a mismanagement of the agricultural economy where Sudan was only able to use a small fraction of its agricultural land;
- b lack of financing as government spending and investments on agriculture and other sectors of the economy were limited and neglected once oil started flowing in 1999 (Siddig, 2012); and
- c lack of expertise in farming, especially in terms of skills and use of modern agricultural technology.

Nevertheless, as a result of South Sudan's secession in 2011, Sudan has lost a large share of its oil revenue: close to 50 per cent of its budget revenue and

75 per cent of its international payments (ADB *et al.*, 2012). Consequently, the agricultural sector is receiving more attention where 'land and water' rich Sudan is seeking to attract the petro-dollars from water-scarce GCC countries (Woertz, 2011) and other investors such as China, Brazil and Egypt through infrastructure development, agricultural investments and trade opportunities.

Parallel to the geopolitical changes in all parts of Sudan and the country's geographic division, a new wave of investments was initiated, globally driven by the 2008 global food crisis (Allan *et al.*, 2013; ECSSR, 2013). The crisis led to temporary export restrictions by food exporters such as Argentina, Russia, India and Vietnam (Woertz, 2013b), where prices towards the end of 2007 reached their highest levels since 1845 – in inflation-adjusted terms – according to the *Economist* magazine's index (Berger and Weber, 2011). In 2011, the price records of 2008 were further broken, leading to a global commodity market shock. These global events led the GCC countries to realize 'the risk of not securing enough food imports at *any* price even if their pockets are lined with petrodollars leaving an immense psychological impact' (Woertz, 2011: 120). Indeed, scarce water resources in the Arab World and limited opportunities for expansion of arable land are often perceived as key obstacles for domestic food production. During the period between 1995 and 2005, the amount of permanent and arable cropland in Arab countries, excluding Sudan, only increased at an annual growth rate of 1.7 per cent compared to 2.3 per cent worldwide and 6.7 per cent in Sudan (Lampietti *et al.*, 2011). These figures reflect the diminished availability of renewable water resources per capita forecasted to be cut in half by 2050 (Akhtar, 2011). Climate change uncertainties bring along a predicted shift in rainfall patterns adding further pressure and stress on water resources. With the Arab region importing 10 per cent of 65–70 million tonnes of cereals from the international reserve globally (ESCWA, 2015), GCC countries are positioning food security as a national priority of utmost importance in terms of 'sovereignty'. The invitation of President Bashir for Arab countries to invest in Sudan's 'land and water resources' fits nicely with the GCC water scarcity and food security narratives, and serves in theory Sudan's shortage of capital in agricultural investments, as well as its domestic food needs. This is especially true if viewed from a Pan-Arab lens led by the LAS, notably among financially strong GCC countries, on the one hand, and human/natural resource-rich Arab and African countries, such as Sudan, Egypt, Ethiopia and Kenya, on the other.

It should be noted, however, that the idea of Arab integration and joint investments is not a new one given its roots since the 1950s and 1960s, under the leadership of Nasser and the Pan-Arab movement, a vision that didn't materialize the way in which it was envisaged at its outset for different reasons and historical factors. Today, almost 60 years after the Pan Arab movement, water-scarce GCC countries have food security high on their agenda, stimulated by global drivers and Pan-Arab regional cooperation narratives. This strategic orientation has been adopted by countries such as Qatar, Saudi Arabia, Kuwait and the UAE, best manifested by their clearly articulated policy objective 'to promote international agricultural investments abroad' in order to minimize their reliance on food imports (ECSSR, 2013). Examples include the Strategy for Sustainable Arab Agricultural Development (2005–2025) approved during the Riyadh Arab Summit in April 2008, as part of the Joint Strategy for Arab Social and Economic Action supported by the Arab Organization for Agricultural Development (AOAD) and the LAS (AOAD, 2007, 2008). This strategy is also complemented by national strategies such as the Qatar National Food Security Programme (QNFSP) and the King Abdullah Initiative for Saudi Agricultural Investment Abroad (KAISAIA).

Institutional landscape: a multitude of actors and stakeholders

State and non-state actors involved in large-scale land and water investments have different profiles and motivations. Several investors from different countries are interested in the Nile Basin's land and water resources, not only from the water-scarce GCC countries, but also from other Arab, Asian as well as BRIC countries (Brazil, Russia, India and China). While GCC states are primarily concerned with food security, private-sector companies from China and Brazil are exploring opportunities in infrastructure development and biofuels as part of the increased BRIC ties with African countries. Examples include Chinese companies with expertise in hydropower such as Synohydro and the China International Company on Water and Electricity, as well as financial institutions such as the China Export Import Bank, involved in providing credit and financing for joint projects and infrastructure investments in Sudan. Brazilian companies are mainly interested in opportunities for biofuel production. Investors involved in large-scale agricultural schemes include sovereign wealth funds (SWFs), state-owned companies or the private sector (Woertz, 2013b: 91).

In terms of GCC land and water-related investment, a multitude of state and non-state actors are involved. Qatar and Saudi Arabia have the most institutionalized approach through the QNFSP and the KAISAIA; Kuwait adopts a top-down approach with the involvement of the ministries of finance and agriculture led by portfolio investors such as the Kuwait Investment Authority tasked with direct investments despite their limited experience in the sector, and the UAE relies more on the private sector and state-owned funds (Woertz, 2013b: 91). Examples of GCC state-owned companies include Qatari 'Hassad Food'.[11] Established in 2008 as a wholly owned subsidiary of the Qatar Investment Authority (QIA), the SWFs of Qatar, its mandate is to run a profitable business with sustainable growth to contribute to the national food security programme (Hassad Food, 2015). The state-owned company is a leading player in foreign agro-investments as the private sector plays a minor role for lack of capacity, with agricultural projects targeting commercial viability on a standalone basis, compliance with international regulations and standards, and a long-term vision not to run the farms itself, but only to issue off-take agreements (Woertz, 2011,

2013b). Despite the fact that the company's website does not mention Sudan as an investment destination, reports indicate that Hassad targets wheat and livestock in Sudan. Another example is the KAISAIA. Launched in January 2009, the KAISAIA was conceptualized as a public–private partnership with a mandate to facilitate land leases, investment opportunities, and with a government role to coordinate different segments of the bureaucracy for the Saudi private sector via political support and co-financing. In April 2009, the Saudi Company for Agricultural Investment and Animal Production, a public investment fund, was also established as part of the KAISAIA with a capital investment of $800 million (Woertz, 2013b: 92). As for the UAE, foreign land acquisitions are more informal where there is no centralized governmental task force or institution with authority comparable to that of the QNFSP. For example, existing development funds such as the AFESD, the state-owned ADFD or the commercially motivated Dubai World subsidiary for investment in natural resources adopt an equity-oriented strategic approach that goes beyond dispersed loan financing, and are often assigned a role to develop large projects in Sudan (Woertz, 2013b: 92).

The AOAD and the Arab Authority for Agricultural Investment and Development (AAAID) are existing institutions since 1972 and 1976, respectively, mandated to ensure Arab food security and support overseas agricultural investments (AAAID, 2013). With a relatively inactive role, the AAAID's board commissioned a study in 2008 to identify the major challenges facing agricultural investments in the Arab countries and recommendations to overcome these. In 2012, the AAAID issued the 'Arab Agricultural Investments Map for Large Scale Projects' identifying three scenarios for agricultural development in the Arab world (AAAID, 2012), including scenario 1: allocating additional land resources to the irrigation sector; scenario 2: using existing comparative advantages in the allocation of agricultural investments to meet basic needs; and scenario 3: increasing agricultural productivity. The investment map identifies potential agricultural opportunities in different countries with a primary focus on Sudan, Egypt, Morocco and Iraq.

Beyond the Arab food security narrative, Sudan is also promoting agricultural investments via COMESA, the Intergovernmental Authority on Development (IGAD) in Eastern Africa and the NBI. Sudan has also become the twelfth member in the COMESA region to sign the Comprehensive Africa Agriculture Development Program (CAADP) compact during a high-level meeting held on 30 July 2013 in Khartoum. Within the CAADP framework, agriculture is a key driver for economic growth in Sudan whereby the government is committed to providing political and economic support through sound agricultural policies supported by regional institutions such as the African Union and the New Partnership for Africa's Development (NEPAD) to assist the country in meeting its commitments and mobilize stakeholders to support the process (CAADP, 2013). Through these various initiatives, Sudanese officials seek to promote a narrative founded on the fact that their country is endowed with enough land and water resources to play a key role in addressing challenges of food security and nutrition in Africa and globally. This narrative may hold true in essence, but is questionable in terms of Sudan's water shares given its transboundary commitments!

The demand side of LSLAs

From the demand side, it is important to acknowledge that LSLAs and land grabs are not homogeneous. LSLAs are driven by different motivations (biofuel/food production/commercial speculation) and often serve larger social, political and economic objectives of domestic and foreign investors. Acquisitions differ in terms of land size, source of water (rain-fed/surface/underground/green water), technology and environmental footprint, among other factors. Beyond the literature rush and the numerous media reports documenting land grabs, some initiatives are continuously attempting to update their databases and provide up-to-date data about LSLAs and their associated characteristics. For instance, the Land Matrix established in April 2012 and advertised as 'a global independent land monitoring initiative' is coordinated by different partners such as the International Land Coalition (ILC), the French-based Centre de Coopération Internationale en Recherche Agronomique pour le Développement (CIRAD), Bern's Centre for Development and Environment, the German Institute for Global and Area Studies (GIGA) and the Deutsche Gesellschaft für Internationale Zusammenarbeit (GIZ), among other regional focal points (Landmatrix, 2015). The Landmatrix database identifies 37 projects in Sudan, including investments by GCC countries such as Saudi Arabia, Kuwait, the UAE and Qatar, in addition to Egypt, Jordan, Lebanon, Turkey, Brazil, Iran, India and Sudan itself. Land size varies greatly between one deal and another ranging from 10,000 to 100,000ha with a few exceptions exceeding these. Not all documented deals have been finalized as indicated in the database by their 'status of negotiation'; some deals are concluded with the contracts signed, others are just intended with an expression of interest, and some deals failed during the negotiation stage. But even with all the confirmed signed deals, the level of implementation on the ground is significantly low, indicating a significant gap, thus 'underlying the very problem it had set out to resolve: the unreliability of media reports' (Woertz, 2013a). In another database developed by a consultant working on large-scale agricultural investments in Sudan, the intended size of GCC land investments is 295,359ha, out of which only 16,200ha were in production in 2014 (Hillhorst, 2015). Other databases documenting LSLAs include GRAIN, a small international nonprofit organization supporting small farmers and social movements towards biodiversity-based food systems also enlisting 19 land deals in Sudan involving different countries (GRAIN, 2012). Overall, in light of the delayed materialization of agricultural investments, quantifying the actual use of 'water and land resources' and their environmental implications remains a challenge given the obscurity of the deals and the lack of quantitative data regarding the actual agricultural production, water withdrawals and their associated aspects (type of crops, water source, productivity, technology, employment effects and exports of agricultural outputs). Nonetheless, the following section presents a few examples of large-scale agricultural projects portraying the diversity of investment modalities from the demand side.

Modalities of water grabs and land investments

Domestic investors play an active role in agricultural investments in Sudan. The Sudanese DAL Agricultural Services manages large-scale farms in the capital Khartoum, and is expanding into other parts of the country using modern agribusiness models, farming techniques and the latest dairy technologies. Al Waha farm is one of its largest projects using centre pivot irrigation, spraying out water and fertilizer in electronically controlled circles having the advantage of efficient water utilization. The farms are set in circles, where each circle comprises 145 feddans, with a total of 92 centre pivot irrigation units, and achieved maximum capacity as planned by the end of 2014. According to an interview conducted with engineer Ali Al Sheikh, director of DAL Agricultural Services, the company's investments have three main purposes: ensuring food security for Sudan, providing fodder for milk supply, and securing hard currency for the country's economy (DAL, 2010, 2014). The company is also seeking to promote milk production in Sudan through training small farmers and providing them with fodder to help improve milk productivity, and buying the milk from small farmers at a fair market price. DAL produces a range of crops such as wheat maize and potatoes for both local consumption and exports.

The KSC is also a key player in large-scale agricultural investments. According to the company's website, Kenana's vision is 'to achieve food security to the Arab world and beyond' with a track record of developing 1 million acres into agricultural production. The KSC production site is located 270km south of Khartoum, where the sugar factory provides 'a model of agro-processing industry in a rural setting' with investments in sugar, food, agricultural products, biofuels, engineering, manufacturing and sustainable integrated solutions for agribusiness (KSC, 2011). The government of Sudan holds 35.33 per cent of the total shares of KSC, while other shareholders include the Kuwait Investment Authority (30.64 per cent), Saudi Arabia (10.97 per cent) and the AAAID (5.59 per cent) (see Figure 7.2). KSC investments also include the White Nile Sugar Company (WNSC), the Mahaseel Agricultural Investment Fund and the Kenana 2020 Projects. The WNSC operation was commissioned on 5 April 2012, including a range of shareholders as shown in Figure 7.3, with a total investment cost of US$1.1 billion. The WNSC follows the KSC's model with the objective of becoming one of the lowest-cost sugar producers in the world. The farm total area is 165,000 acres located on the eastern bank of the White Nile, growing cane for manufacturing with a targeted production capacity of 24,000 metric tonnes (MT)/cane/day, 450,000 tonnes of sugar, 128MW of electricity, 60 million litres of ethanol and 100,000MT of animal feed. The Mahaseel Agricultural Investment Fund is a private equity investment involved

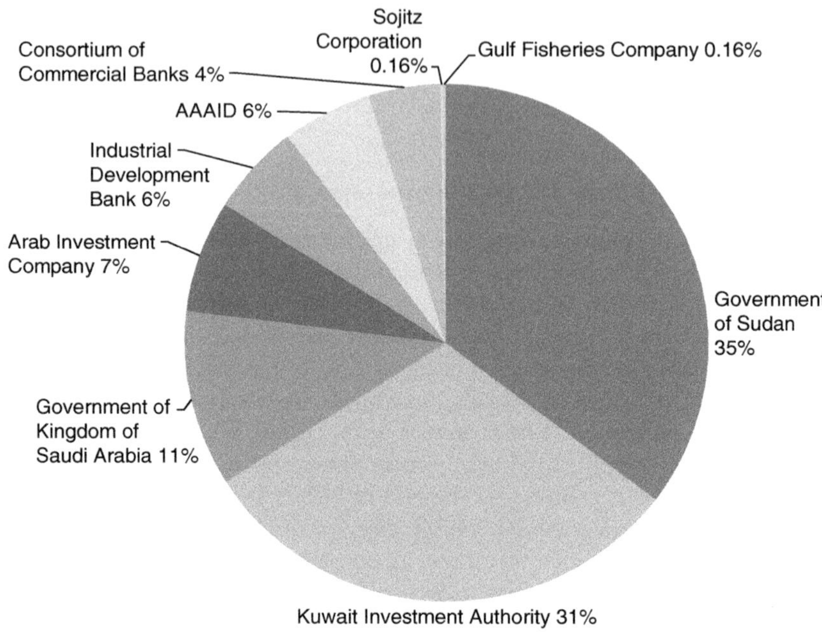

Figure 7.2 KSC limited shareholders

Source: Kenana Sugar Company, www.kenana.com (accessed 17 January 2015).

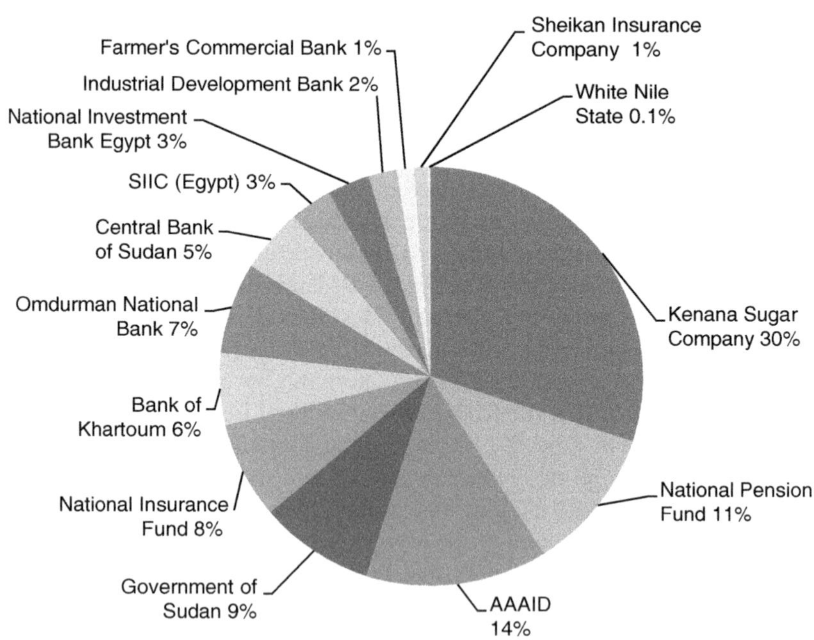

Figure 7.3 Shareholders in the WNSC

Source: Kenana Sugar Company, www.kenana.com (accessed 17 January 2015).

in agriculture and agro-food related industries. The legal structure of the fund is a Cayman Islands exempted limited partnership with aggregate capital commitments of US$1 billion. Kenana 2020 Projects include agriculture activities, such as the Redais Sugar Project (Kenana II), and the Kenana Red Sea Sugar Refinery Project. In the energy sector, projects include ethanol expansion, biodiesel and bio-waste energy projects. For animal production, projects include animal feed and aqua farming projects. Other projects include logistics via the Red Sea Terminal Project, engineering activities via the Kenana Equipment Manufacturing Expansion Project, as well as consumer products and food ingredients projects including liquid sugar, vegetable oil, horticulture, paper and packaging.

Other modalities of investments also exist. For example, according to the Arab Brasilian Chamber of Commerce, during the Brazil–Africa integration seminar conducted on 29 August 2014 in Ceará–Fortaleza, two projects involving Brazilians and Arabs were presented involving investment by the Algerian company Cevital in Brazil, while the other entailed the Brazilian Agro-Industrial Company in Sudan (Farmlandgrab, 2014a). The chairman of the company, Paulo Hegg, presented his company's ongoing agricultural projects in Sudan, including growing grains and cotton, where 'the company's planted area has increased from 7,300 hectares in 2011 to 11,000 hectares in 2014 with projected revenues to reach US$22 million' (ANBA, 2014).

In terms of Gulf (GCC) foreign investments in Sudan, the Economist Intelligence Unit indicates that Sudan's National Investment Board issued a report in February 2015 claiming that UAE investments in Sudan amount to US$6.7 billion, out of which Emirati investments in agricultural schemes are US$5.7 billion in total, with 19 projects (ten in Khartoum state and nine in the central states of Gedaref, Nile, Northern and White Nile) funded between 2001 and 2013 (EIU, 2015). The EIU questions the accuracy of the reports, stating that 'claims about investments are often inaccurate and authorities do not have strong records when it comes to monitoring and reporting foreign and domestic investments'. Saudi Arabian investments are also targeting Sudan. Four Saudi companies recently acquired agricultural lands covering an area of approximately 4,000 acres in Sudan's northern region (*Arab News*, 2014). In January 2014, Ibrahim Al-Khidir, governor of the Northern region, announced the entry of Saudi companies (Al-Safi, Al-Marai, Tabuk Agriculture Co and Al-Jouf Co) in the field of agriculture. According to the governor, the Northern region has attracted 32 investment projects from Gulf countries, covering an approximate area of 14 million acres. These are all located in the upper areas of the River Nile, thus benefiting from the fertile land and available water resources (surface and underground) in the Nubian Basin and the available infrastructure in the Northern region, including the Dongola Airport, a road network facilitating access to markets and a recently approved installation of a power transformer station with Arab financing (*Arab News*, 2014). Other operating Saudi investors include Al-Rajhi's Al Kafa2a project established near Barabar City in the state of the Nile River in the

Northern region. The Al Kafa2a project is planned on 5,000ha relying on a main canal 5km from the Nile to the external periphery of the project, which branches out to smaller canals totalling 35km. Currently, 75 axes of the 100 planned are already cultivated, including wheat, corn and alfalfa, using modern technology and agricultural mechanization (Vimeo, 2014). Plans for expansion include an additional two areas of 50,000 and 30,000 acres to the eastern and southern parts of the project, respectively. In its reports, the company identifies its initial challenges to be the choice of the irrigation system, taking into consideration soil and temperature characteristics. In the River Nile State, the Qatari Hassad Food project in Abu Hamad is a US$205 million agricultural investment, supported by a 310km power line, which includes four transformer stations, Berber, Al Ghbash, Abu Hamad and Hassad, inaugurated on 23 June 2013 by Sudan's president and a Qatari delegation (*Gulf Times*, 2013).

Financial investors and private equity firms are also involved in land and water investments. A good example under this category of investments is GLB Invest SAL, a Lebanese offshore company operating in the trade of agricultural raw materials. Initiated in 2011, GLB Invest acquired 87,200ha and is developing a centre-pivot irrigated alfalfa farm in North Sudan (Farmlandgrab, 2014b; GLB Invest, 2014). The first phase has already been completed whereby the project is planned over five phases ending in 2018, with planned production capacity to reach 40,000MT in 2014, to be sold primarily to the UAE and Saudi Arabia. In December 2013, Bemo Securitisation (BSEC) and Bemo Saudi Fransi Finance (BSFF) acted as joint sell-side advisers to GLB Invest SAL and facilitated an acquisition of a minority stake by M1 Harvest in a deal worth US$16.5 million. Firas Badra, the chairman of GLB Invest SAL, had declared in May 2013 during an investment conference in Khartoum that his investments in this project will reach US$800 million by 2019 (*Business News*, 2014). M1 Harvest is a subsidiary of M1 Group, a diversified investment group based in Lebanon and owned by Prime Minister Mikati's family. M1 Group is active in telecom, real estate, banking, fashion and aviation, among other industries (*Business News*, 2014). Further to this deal, it was announced in November 2014 that a memorandum of understanding was signed in Khartoum at the premises of the Ministry of Finance and National Economy with the Fras Badra F. B. Green Company for the establishment of an investment fund for the agricultural projects in Sudan as a new way of raising financial capital to mobilize natural and agricultural resources through the implementation of a Five-Year Programme using modern technology as stated by the minister of finance (Farmlandgrab, 2014b).

These recent deals were also preceded by other investments. In the White Nile State, Citadel Capital, Egypt's largest private equity company, launched the Wafra Fund in 2007 as a platform to invest in the Sudanese agriculture sector via a portfolio of companies such as the Concord Irrigated Crops in Unity, previously known as the Sudanese Egyptian Agricultural Crops Company (SEAC) (250,000 feddans/105,000ha in Southern Sudan), Sabina (254,000 feddans/106,680ha in Northern Sudan) and Al-Nahda for Integrated

Solutions (EJATLAS, 2014a). The Al-Nahda project for rice production in Ed Dueim (150km south of Khartoum) is Sudan's first large-scale commercial rice farm planned on 25,210ha (60,000 feddans), and rice will be processed on a site-based mill. The project's objective is to grow rice primarily for domestic consumption in Sudan, a net importer of rice, with the excess production to be exported to Africa and the Middle East. Land is flooded for up to eight months per year by water held back by the Jebel Al-Awliaa Dam, where earthworks for the farm include the construction of a major levee bank to exclude the water regulated back onto the property (Farmlandgrab, 2010). It is worth noting that in November 2011, the US Government's Overseas Private Investment Corporation provided Citadel with a US$150 million loan package to help expand its subsidiaries, US$115 million of which was earmarked for crop production in South Sudan (EJATLAS, 2014a).

Challenges facing investors

Despite this growing interest, the logic of land acquisitions has been challenged by the former Saudi chief trade negotiator Fawaz al-Alami, who has argued against foreign agro-investments, indicating that 'they will not guarantee food security in times of crisis' when exporters are most likely to resort to export restrictions, and instead suggested the establishment of strategic food reserves (Woertz, 2011). Other investors such as Eisa Al Ghurair, chairman of Al Ghurair Resources,[12] the agricultural commodities arm of Dubai-based Al Ghurair Investment, also warned investors against buying farmland in foreign countries, citing political risk. He indicated during a press conference at Gulfood 2015 in Dubai that the idea of foreign investors buying up farmland has become a 'touchy subject' with the farmers and government implying considerable political risk in buying up farmland in certain countries with unstable governments (Farmlandgrab, 2015). Examples include the protests against the Gulf-financed Merowe Dam in 2006, as well as the Kajbar and Dal Dams in the north of Sudan and the Kuwaiti-financed heightening of the Roseires Dam (Woertz, 2013a), which are sometimes perceived as 'part of an official plan of demographic engineering' (Committee of Anti Dal-Kajbar Dams, 2011). Other protests in Sudan occurred at Egyptian and Qatari project sites in Sennar, where locals complained about lacking job opportunities in the capital-intensive projects, similar to protests in the same state leading to a collapsed deal by Saudi businessmen, in addition to North Kordofan pastoralists threatening violence against a Saudi cattle ranching project on their ancestral land.

Additional challenges include the lack and inadequacy of roads. Sudan's agricultural lands and sizable parts of its economic productive areas are isolated from the markets. It is estimated that Sudan needs 2,900km of roads to meet regional connectivity standards (linking Khartoum to international frontiers); 5,300km to meet national connectivity standards (linking all provincial capitals to the regional network); and a further 34,201km to meet rural connectivity standards (linking land responsible for 80 per cent of existing agricultural production value to the national network, as well as linking land with the capability of producing 50 per cent of the non-realized agricultural value) (Ranganathan and Briceño-Garmendia, 2011). As such, road development is a necessary precondition to exploring Sudan's agricultural potential. Furthermore, according to the AAAID (2012: 16), the slow materialization of agricultural investments in Sudan is due to some factors and challenges such as:

1 land allocation, tenure and registration issues, especially as it relates to renewing rents and government withdrawing parts of the allocated land areas from investors;
2 non-compliance with tax and duties exemption as stated in the agreement resulting in higher costs for the investor;
3 overlap and clashing mandates between different government departments and institutions on multiple scales, such as federal ministries, state ministries and local government offices;
4 lack of economic and political stability;
5 weak labour laws and human resources constraints;
6 inconsistent fees and taxes – random increase of additional fees and new taxation measures.

Overall, weak infrastructure, weak institutions and lack of enabling environments both economically and politically are major constraints facing the flourishing of agricultural investments in a country such as Sudan. These are similar constraints to the ones faced by Sudan in the unachieved 'breadbasket' vision of the 1970s.

Other factors that should be considered when buying farmland abroad include the country's logistics, investment environment and population growth. Sudan's investment policies seek to facilitate market access and the creation of an enabling trade environment for foreign investors. During a visit to the capital, Riyadh, Sudan's investment minister announced that his government will set up six free zones[13] for Saudi investors, offering them access to the markets of eastern and southern Africa (EIU, 2009). He also claimed that changes to Sudan's investment law had addressed '37 obstacles' that previously hampered investments, where constitutional amendments had transferred some of the power to grant land for investment from Sudan's provincial (state) authorities to the president (EIU, 2009). A change claimed to be a measure to prevent landownership disputes, as happened in 2013 when the residents of a Khartoum suburb 'Om Doum' protested against plans to grant local land to a Saudi investor. Yet, it is not uncommon for government officials to make large claims about development plans with no actual follow-up, whereby governments often 'fail to garner firm commitments of money from Saudi Arabia and the Gulf, primarily because the announced plans were sketchy and overambitious' (EIU, 2009).

Transboundary water resources governance and Nile hydropolitics: empirical and conceptual implications

The study and research of transboundary water resources management (TWM) has been typically addressed from a *state-centric* lens. The emergence of new players with varying degrees of independence from their respective governments including state enterprises and non-state actors such as SWFs, state-owned companies, the private sector, including specialized agricultural companies, and private equity firms, points to a different pattern of investments in land and water resources. In this respect, the topic of land acquisition and transboundary water management is unchartered territory where the effect that 'large-scale land-water related investments' will have on transboundary water resources has not yet been analysed (Jägerskog *et al.*, 2012). Therefore, understanding the role of 'state-corporate-capital alliances' in the appropriation of natural (land and water) resources is essential given the wider 'battle of ideas' regarding water security (Lankford *et al.*, 2013), notably on a transboundary scale.

For instance, from a hydropolitical standpoint, the different implications of these investments are uncertain. It is not understood whether or not investments by non-state actors are going to increase the political tensions between upstream and downstream countries due to increased water withdrawals beyond the allocated legal shares in a shared river basin. Similarly, it is not clear if these investments are going to result in an increased use of water resources with negative environmental footprints on a river basin scale leading to catastrophic implications downstream within a country and consequently across the borders. Prioritization of different productive and economic uses of water resources within and across countries for hydropower generation, agricultural production, tourism, transportation or other uses is also not evident. How are these issues going to be resolved with the increasing private investments? Who decides what? In whose interest? These are all unanswered questions. Other sides of the argument may view positive implications of these investments on food and energy security on a regional scale, despite their negative implications on livelihoods and local communities. Furthermore, a key issue in the relationship between LSLAs and transboundary water resources is the fact that associated use of water resources is rarely captured or quantified. Therefore, it is necessary to explore possible tools to quantify and provide projections regarding the amounts of water withdrawals resulting from the different large-scale agricultural investments. Some efforts are currently under way such as the 'Water Accounting Tool' (Bastiaanssen *et al.*, 2014); yet the lack of data due to political sensitivities is a main obstacle.

In terms of Nile Basin hydropolitics, Egypt and Sudan have historically formed an alliance to maintain the hydropolitical status quo. This alliance could become fragile if Sudan develops its water infrastructure for agriculture and uses more than its water quota under the existing agreements. According to the minister of water resources, the government in Sudan developed a ten-year plan known as the Comprehensive National Strategy, 1992–2002, aimed at increased food security, sustained agricultural development, efficient resource utilization and yield enhancement. The Sudanese government had also estimated that the country will need 32bcm of water by 2025 to meet food security and other needs (Taha, 2010: 207–209). The situation can be further complicated taking into account the fact that water shares between Sudan and South Sudan are not sorted out and the future of the 18.5 billion annual water flow is still undecided. Added to this are the hydropolitical contestations and negotiations between upstream and downstream countries regarding historical water shares based on the 1959 Agreement. In this respect, Egypt's concerns are not strictly limited to the lack of benefiting from the surplus previously coming from Sudan, but also by the potential risks on its water security. Yet, Egyptian investments are present in Sudan. Egyptian and Sudanese officials indeed discussed wheat production in North Sudan on 2 million feddans in 2011, and the government of Sudan had also allocated 100,000 acres (40,000ha) divided into small slots of 10 acres (4ha) each to small-scale Egyptian farmers as announced by the Sudanese minister of investment in Khartoum in December 2014 confirming that 'Sudan gives maximum priority to Egyptian investors in the country' (Farmlandgrab, 2014c). This approach can be viewed as a way of compromise with Khartoum, which could perhaps represent a good example of the 'co-existence of conflict and cooperation' in TWM (Mirumachi and Allan, 2007; Earle *et al.*, 2010), albeit from a state-centric standpoint.

Furthermore, given the perceived abundance of land and water resources in the Nile Basin, different state and non-state actors have established new governance mechanisms to cope with the requirements of strategic agro-investments beyond the traditional role of the state. Gulf States such as Qatar and Saudi Arabia have created new institutions to coordinate food security policies and agricultural investments between various state agencies and the private sector. In this respect, Arab states are 'dominant in formulating foreign agro-investment strategies and try to pave the way for SWFs, state-owned companies, or the private sector by negotiating framework agreements' (Woertz, 2013b: 51). As a result, the successful completion of these investments in Sudan and other Nile Basin countries (e.g. Ethiopia, Kenya, etc.) could put the Gulf States in confrontation with Egypt. As such, there is a lack of clarity about the impact of large-scale agricultural investments upon *existing state-centric multilateral organizations* and governance structures primarily mandated with questions of TWM. A primary example in the context of this research is the NBI, which represents a state-centric intergovernmental institution supported through a donor functionalist approach, mandated to promote peace and security through transboundary water resources cooperation (Nile Basin Initiative, 2014). Yet, with the growing influence and power of state and non-state actors, including private investors 'from within and outside the basin' and their associated changing patterns of land and water use, it is unclear how an institution such as the NBI can maintain its role. It is also unclear how it will influence the hydropolitical state-centric relations in light of the increasing power of the exogenous players endowed with capital and technology, actively appropriating large amounts of land and water resources for private interests most likely from outside the basin.

Conceptual implication: state–capital alliances – towards private water governance?

The study of political economy outlines the interaction of economic and political systems, including both ways in which political beliefs can affect the distribution of economic resources, and how economic interests can influence the political activities of governments (Evans, 2011: 31; Araral and Wang, 2013). 'A political economy account goes further, however, in asking why some coalitions of interest prevail over others' (Newell, 2013: 43). In retrospect, the institutions, practices and ideology represent a structure of environmental governance (Finger, 2005: 284–285), where the relationship between economy and politics is fundamental (Clapp and Dauvergne, 2005). Therefore, there is a need to understand global economic structures, processes and the power relations between them as well as the environmental implications of the actual everyday functioning of the markets (Clapp and Helleiner, 2012a, 2012b). It should be noted, however, that conventional studies of international environmental politics have viewed global governance as a function provided primarily by the state systems through processes of international negotiations and regime building (Falkner, 2003: 75). Contradictory to this view that state systems dominate global governance, Conca (2006) suggests that threats to the world's water commons have resisted the establishment of effective global or interstate water agreements through intergovernmental bargaining because the conditions for successful interstate cooperation cannot be imposed upon water controversies (Araral and Wang, 2013: 3953). However, he argues, while interstate water diplomacy has faltered, less formalized institutions – socially and politically embedded rules, roles and practices – have emerged to help shape water governance locally and globally (Aral and Wang, 2013: 3953).

Applying such views to the recurring debates on water security, neo-liberal capitalism and the process of globalization, LSLAs and water grabbing practices could be perceived as facilitating the transfer of power and authority from the public to the private sector (Falkner, 2003: 73). The 'privatization' of global environmental politics could be therefore regarded as a process that undermines the established, 'inefficient' state-centric models of global environmentalism (Falkner, 2003: 81). The increased participation of state and non-state actors in acquiring natural (land and water) resources abroad may eventually head towards the notion of 'financialization' of natural resources supported by global commodities market dynamics (Clapp and Helleiner, 2012a: 498).

Conclusion

Hydropolitics of the Arab food basket: Egypt and Sudan at the intersection of Arab food security and African hydro-diplomacy

Historically, the Sudan 'breadbasket vision' was not achieved. As a key country in the Nile Basin, Sudan is seeking to revive this vision to make up for its lost oil revenue due to South Sudan's secession in 2011. The government has embarked on a massive infrastructure and dam development programme for hydropower and irrigation using Chinese expertise and credit, as well as Arab financing. Beyond infrastructure development, Sudan is also seeking to promote an ambitious agricultural investment plan as announced by the Ministry of Investment in 2013 supplying large-scale agricultural lands and water resources. In this sense, Sudan is positioning itself as a primary agricultural investment destination due to its abundant land and water resources, often viewed as a cornerstone for a joint Arab food security strategy.

The intersection of Arab water scarcity and food security narratives on the one hand, and the Nile Basin hydropolitics on the other, is therefore best manifested by GCC and Arab agricultural investments in Sudan. Arab investors and water-scarce GCC countries are seeking to 'promote agricultural investments abroad' as part of their food security strategy in order to ensure their food 'sovereignty', especially following the 2008 and 2011 global food crisis. While Qatar and Saudi Arabia have established new institutions addressing the question of food security and agricultural investments abroad, other investors from the UAE and Kuwait do not have a comparable degree of institutionalization, but strongly support the participation of the private sector in agricultural investments abroad (Woertz, 2013b). Other financial investors and private sector companies from Brazil, Lebanon and Jordan targeting biofuel production and agricultural commodities trade are also involved in LSLAs abroad primarily for commercial purposes. Domestic investors are also active in LSI schemes under different investment modalities. On the demand side, different actors and a range of large-scale agricultural investment modalities exist in Sudan. These include state-owned enterprises, SWFs, specialized agricultural private sector companies and private equity investors. This chapter sought examples for these different categories of investments by providing evidence about the growing role and agency of different state and non-state actors involved in the acquisition of land and water resources. From the demand side, the classification of LSLAs could be proposed as investments by:

- GCC and Arab countries for food security purposes: investments are undertaken by GCC governments for food production and export to home countries within the framework of the Arab economic cooperation.
- GCC and Arab investors for trading, export and commercial purposes in agricultural commodities to the larger MENA/GCC regions.
- BRIC countries, notably Brazil, primarily targeting biofuel production.

- Other non-state actors and financial investors from Arab and Asian countries for commercial and exports: investors are from South Korea, Egypt, Lebanon, Turkey and Jordan, among others.

In terms of implications on transboundary water resources governance and Nile Basin hydropolitics, the chapter argues that the role of state and non-state actors in water grabbing is often missing from transboundary water resources governance and is a new element in the Nile Basin hydropolitics, which deserves further attention. The different state and non-state actors, especially the private companies, are not formally involved in any institutional arrangements on a river basin scale to address practices related to water and land resources investments. In this respect, LSLAs for agricultural and biofuel production are associated with 'water grabbing' practices where water-scarce countries as well as financial investors seek to grab water resources abroad to serve their national food security needs or financial returns on their investments. As such, the implications of these water grabs by domestic and foreign investors remain unclear on a transboundary scale. Unless agro-investments in Sudan receive proper social and environmental planning, achieving food security for countries such as Saudi Arabia and Qatar may end up coming at the expense of Sudan's and other Nile Basin countries' own food and water security. These emerging elements raise questions around the role of non-state actors and their business practices and how they manage their dependencies and impacts upon food, energy, water and the environment, as well as the implications of these across international borders on a scale of a transboundary river basin. This is especially the case when examining the role of investors from outside the basin, as well as investors from within the basin investing in other countries (i.e. Egypt investing in Sudan and the implications on its water shares downstream). Overall, understanding the interests and roles of Arab, Chinese and Brazilian investments – among others – in water-related infrastructure and LSLAs in Sudan for agricultural/biofuel production, and situating it with the larger hydropolitical context of the Nile River Basin, is not an easy task. Therefore, it is the interaction between actors at various scales that is critical for understanding change and continuity – and inevitable inconsistencies and contradictions – in the political constructions of water (Mosse and Sivan, 2003).

Notes

1 Other treaties took place prior to the 1929 Agreement such as the 1902 Treaty between Ethiopia and Britain on behalf of Egypt marking the 'principle of non-interference with the flow of the Blue Nile', confirmed by the Tripartite Agreement of 1906 among the colonial powers of Britain, France and Italy, leading to an exchange of notes in 1925 where Italy recognized the 'prior hydraulic rights' of Egypt and Sudan and agreed not to construct any works likely to modify the flow of the Ethiopian tributaries of the Nile (Shapland, 1997: 70).

2 Sudan was planning to build two dams that led to the development of the 1929 Treaty: Sennar Dam completed in July 1925 primarily for cotton irrigation, and Jabal Awliya Dam, which was being planned prior to the 1929 Agreement but was completed in 1936 for hydropower and irrigation (Awulachew *et al.*, 2012).
3 The Roseires Dam in 1950 triggered Egypt's protests, leading to the 1959 Agreement with Sudan.
4 For nomenclature, *all parts of Sudan* refers to Sudan prior to the split of the country, *Sudan* refers to the northern part of Sudan, and *South Sudan* is the newly independent country.
5 Sudan lost almost one-third of its land size from 2.5 million km^2 to 1.86 million km^2.
6 Water shares were not part of the signed independence agreement in 2011 and were not addressed during political settlement.
7 Cotton was a mandatory crop, but in the 1970s it was partly replaced by sugar cane.
8 The project is supervised by different government departments such as the Dam Implementation Unit, the National Corporation for Antiquities and Museums, the Ministry of Irrigation and Water Resources, the Ministry of Electricity and the Sudanese National Council for Water Resources.
9 The site of the twin dam is located 20km (12 miles) upstream from the junction of the Atbarah and Setit Rivers and about 80km (50 miles) to the south of the Khashm el-Girba Dam.
10 Cotton, sorghum grasses, sunflower and groundnut, as well as animal husbandry.
11 Hassad Food has projects in Australia, and is negotiating deals in Brazil, Pakistan and Georgia for sugar, rice and wheat production.
12 Al Ghurair Resources imports grains from Europe, USA, Canada, Australia and South America. It also operates facilities in Algeria and Egypt.
13 Sudan currently has free zones at Suakin (on the Red Sea coast) and Aljaily, near Khartoum. Under a 1994 law, these are meant to provide exemptions from profit taxes for 15 years, as well as exemptions from customs fees and other taxes.

References

AAAID (Arab Authority for Agricultural Investment and Development) (2012). *Khareta Zera3ya Estethmareya 3arabeya le tanfeez Masharee3 Zera3eya Kobra, Map for Large Scale Arab Agricultural Investment Projects*. Translated from Arabic. Khartoum, Sudan.

AAAID (2013). *Al Taqrir al Thanawy – Annual Report 2013 of the Arab Authority for Agricultural Investment and Development*. Khartoum, Sudan.

Abdel Wahab, A. (2011). *Water Security in the Nile Basin: Controversies of Development and Stability. Al Amn Al Ma2y Fi 7od Al-Nil: Ishkaleyat Al Tanmeya wal Istekrar. Cairo, Egypt.* Cairo: Ahram Center for Political and Strategic Studies.

ADB, OECD and UNDP (2012). *Sudan African Economic Outlook (AEO) 2012*. Available at: www.afdb.org/fileadmin/uploads/afdb/Documents/Publications/Sudan Full PDF Country Note.pdf (accessed 18 March 2015).

ADB, OECD and UNDP (2015). *Sudan African Economic Outlook 2015*, Available at: www.africaneconomicoutlook.org/fileadmin/uploads/aeo/2015/CN_data/CN_Long_EN/Sudan_GB_2015.pdf (accessed 3 May 2015).

Akhtar, S. (2011). *Food Security in the Arab World: Price Volatility and Vulnerabilities and the World Bank Response Global, Vice President, Middle East and North Africa*. World Bank, Key Note Address, First Arab Development Symposium, Food and Water Security in the Arab World. Kuwait.

Allan, J. A., Keulertz, M., Sojamo, S. and Warner, J. (eds) (2013). *Handbook of Land and Water Grabs in Africa: Foreign Direct Investment and Food and Water Security*. London: Routledge.

ANBA (2014). Arabs participate in Brazil-Africa integration. Translated by G. Pomerancblum. *Brasil Arab News Agency*. Available at: http://www2.anba.com.br/noticia/21864760/agribusiness/arabs-participate-in-brazil-africa-integration (accessed 2 April 2015).

AOAD (Arab Organization for Agricultural Development) (2007). *Strategy for Sustainable Arab Agricultural Development for the Upcoming Two Decades (2005–2025)*. Khartoum.

AOAD (2008). *Riyadh Declaration to Enhance Arab Cooperation to Face World Food Crises League of Arab States*. Khartoum.

Arab News (2014). Four local firms acquire Sudan agricultural lands. 24 January. Available at: www.arabnews.com/news/514411 (accessed 2 April 2015).

Araral, E. and Wang, Y. (2013). Water governance 2.0: A review and second generation research agenda. *Water Resources Management*, 27(11): 3945–3957.

Al Bashir, O. (2015). Speech President Al Bashir at opening ceremony of Egypt Economic Development Conference, Sharm El Sheikh.

Awulachew, S. B., Smakhtin, V., Molden, D. and Peden, D. (2012). *The Nile River Basin: Water, Agriculture, Governance and Livelihoods*. London: Routledge. Available at: http://link.springer.com/chapter/10.1007/978-3-319-02720-3_23/fulltext.html.

Bastiaanssen, W., Karimi, P., Rebelo, L.-M., Duan, Z., Senay, G., Muttuwatte, L. and Smakhtin, V. (2014). Earth observation based assessment of the water production and water consumption of Nile Basin agro-ecosystems. *Remote Sensing*, 6(11): 10306–10334. Available at: www.mdpi.com/2072-4292/6/11/10306 (accessed 5 March 2015).

Berger, M. T. and Weber, H. (2011). *War, Peace and Progress in the 21st Century: Development, Violence and Insecurity*. London: Routledge.

Business News (2014). Mikati group invests in Sudan (Leila Rahbani). Available at: www.businessnews.com.lb/cms/Story/StoryDetails.aspx?ItemID=3758 (accessed 27 March 2015).

CAADP (2013). Comprehensive Africa Agriculture Development Program. PRESS RELEASE: Sudan Signs the National CAADP Compact/Release Date: 30 July 2013, pp1–3. Available at: www.comesa.int/index.php?option=com_contentandview=articleandid=829:press-release-sudan-signs-the-national-caadp-compactandcatid=6:press-releasesandItemid=57 (accessed 10 April 2015).

CAPMAS (Central Agency for Public Mobilization and Statistics) (2014a). Central Agency for Public Mobilization and Statistics press release on 'Water Resources and its Rational Use in Egypt'. Available at: http://capmas.gov.eg/pepo/603_e.pdf (accessed 7 April 2014).

CAPMAS (2014b). *State Information System 2015: Population Estimates By Sex and Governorate*, 1 January. Available at: www.sis.gov.eg/En/Templates/Articles/tmpArticles.aspx?CatID=19#.VUNoABcjinc (accessed 1 March 2015).

Clapp, J. and Dauvergne, P. (2005). *Paths to a Green World: The Political Economy of the Global Environment*. Cambridge/London: Cambridge University Press.

Clapp, J. and Helleiner, E. (2012a). International political economy and the environment: Back to the basics? *International Affairs*, 88(3): 485–501.

Clapp, J. and Helleiner, E. (2012b). Troubled futures? The global food crisis and the politics of agricultural derivatives regulation. *Review of International Political Economy*, 19(2): 181–207.

COMESA Regional Investment Agency (RIA) (2013). *Investment Teaser 2012–2013: Investment Projects and Opportunities*. Cairo, Egypt.

Committee of Anti Dal-Kajbar Dams (2011). A letter to the Sinohydro Company, China and the Chinese Government via the Chinese Embassy, Khartoum: A letter of protest and resistance, pp1–33. Available at: www.sudan-forall.org/Anti-Dal-Kajbar-Dams_Executive-Summary.pdf (accessed 5 March 2015).

Conca, K. (2006). *Governing Water: Contentious Transnational Politics and Global Institution Building*. Cambridge, MA: MIT Press.

Dabanga (2014). Sudan allocates 100,000 acres to Egyptian farmers. Available at: www.dabangasudan.org/en/all-news/article/sudan-allocates-100-000-acres-to-egyptian-farmers (accessed 17 February 2015).

DAL (2010). *Dal Group Food Division Corporate Video (English)*. Available at: www.youtube.com/watch?v=MoJ1inY7Vzc (accessed 5 March 2015).

DAL (2014). *Sudan Agriculture*. Available at: www.youtube.com/watch?v=mbjLTft9D-0 (accessed 5 March 2015).

Deininger, K. and Byerlee, D. (2011). *Rising Global Interest in Farmland: Can it Yield Sustainable and Equitable Benefits?* International Bank for Reconstruction and Development/World Bank. Available at: http://siteresources.worldbank.org/INTARD/Resources/ESW_Sept7_final_final.pdf (accessed 16 September 2014).

Earle, A., Jägerskog, A. and Öjendal, J. (2010). *Transboundary Water Management Principles and Practice*. London: Earthscan.

ECSSR (2013). *Water and Food Security in the Arabian Gulf.* Abu Dhabi: Emirates Center for Strategic Studies and Research, Abu Dhabi.

EIU (Economist Intelligence Unit) (2009). *GULF STATES: Saudis Renew Search for Food Security*. Available at: http://search.proquest.com/docview/192457216?accountid=14182 (accessed 9 March 2015).

EIU (2015). *Sudan Economy: Quick View – UAE Loan Agreed*. Available at: http://search.proquest.com/docview/1659818246?accountid=14182 (accessed 9 March 2015).

EJATLAS (2014a). Citadel Capital/Sabina and Al Nahda for integrated solutions irrigated crops in Sudan. *Environmental Juctice Atlas*. Available at: http://ejatlas.org/conflict/citadel-capital-sabina-and-al-nahda-for-integrated-solutions-irrigated-crops-in-white-nile-state-sudan (accessed 7 April 2015).

EJATLAS (2014b). Kajbar Dam hydropower scheme, Sudan. *Environmental Justice Atlas*. Available at: http://ejatlas.org/conflict/kajbar-dam-hydropower-scheme-sudan (accessed 7 April 2015).

Elhance, A. P. (1999). *Hydropolitics in the Third World: Conflict and Cooperation in International River Basins*. Washington, DC: US Institute of Peace Press.

Embassy Press (2015). *Agriculture in Sudan – Embassy of the Republic of Sudan in Washington D.C.* Available at: www.sudanembassy.org/index.php?option=com_contentandview=articleandid=88andItemid=172c (accessed 5 March 2015).

ESCWA (2015). *Conceptual Frameworks for Understanding the Water, Energy and Food Security Nexus*. E/ESCWA/SDPD/2015/WP. 20.

Evans, J. P. (2011). *Environmental Governance*. London: Routledge.

Falkner, R. (2003). Private environmental governance and international relations: Exploring the links. *Global Environmental Politics*, 3(2): 72–87.

Farmlandgrab (2010). Wafra Portfolio Company to build Sudan's first large-scale commercial rice farm. Available at: http://farmlandgrab.org/post/view/11999 (accessed 20 March 2015).

Farmlandgrab (2014a). Arabs participate in Brazil-Africa integration. Available at: http://farmlandgrab.org/post/view/23936-arabs-participate-in-brazil-africa-integration. (accessed 17 February 2015).

Farmlandgrab (2014b). MoU signed for agricultural investment fund. Available at: www.farmlandgrab.org/post/view/24248-mou-signed-for-agricultural-investment-fund (accessed 20 March 2015).

Farmlandgrab (2014c). Sudan allocates 100,000 acres to Egyptian farmers. Available at: http://farmlandgrab.org/post/view/24316-sudan-allocates-100-000-acres-to-egyptian-farmers (accessed 20 March 2015).

Farmlandgrab (2015). Investors should work with farmers to secure food supply: Al Ghurair. Available at: http://farmlandgrab.org/post/view/24505-investors-should-work-with-farmers-to-secure-food-supply-al-ghurair (accessed 11 April 2015).

Finger M. (2005). The new water paradigm: The privatization of governance and the instrumentalization of the state. In D. L. Levy and P. J. Newell (eds) *The Business of Global Environmental Governance*. Boston, MA: MIT Press, pp275–304.

Gareau, B. J. and Crow, B. (2006). Ken Conca, governing water: Contentious transnational politics and global institution building. *International Environmental Agreements: Politics, Law and Economics*, 6(3): 317–320. Available at: http://link.springer.com/10.1007/s10784-006-9007-1 (accessed 6 October 2014).

Geopoliticalmonitor (2013). Sudan's agricultural sector sees investment opportunities. Available at: www.eurasiareview.com/17102013-sudans-agricultural-sector-sees-investment-opportunities-analysis (accessed 22 March 2015).

GLB Invest (2014). GLB invest alfalfa farm Sudan. Available at: www.youtube.com/watch?v=L8PC171MyMQandindex=67andlist=PL9C410C53273FA514 (accessed 5 March 2015).

GRAIN (2012). GRAIN/Land grab deals. January, p62. Available at: www.grain.org (accessed 16 September 2014).

Gulf Times (2013). Qatari delegation in Sudan for inauguration of Hassad Food Company's project. Available at: www.gulf-times.com/qatar/178/details/357175/qatari-delegation-in-sudan-for-inauguration-of-hassad-food-company%E2%80%99s-project (accessed 7 April 2015).

Hassad Food (2015). Hassad Food. Available at: www.hassad.com (accessed 20 February 2015).

Hillhorst, B. (2015). *Data Base: GCC Agricultural Investments in Sudan*. Factsheet. Primary data from interviews/data from Bart Hillhorst.

IIED (2011). *Are Land Deals Driving 'Water Grabs'? Briefing: The Global Land Rush*. London: IIED.

International Rivers (2011a). *Merowe Dam*. Available at: www.internationalrivers.org/campaigns/merowe-dam-sudan-0 (accessed 20 February 2015).

International Rivers (2011b). *The Kajbar Dam, Sudan*. Available at: www.internationalrivers.org/campaigns/kajbar-dam-sudan (accessed 20 February 2015).

Jägerskog, A., Cascão, A., Hårsmar, M. and Kim, K. (2012). *Land Acquisition: How Will it Affect Transboundary Waters?* Report No. 30, Stockholm: SIWI.

KSC (2011). *Kenana Sugar Company*. Available at: www.kenana.com (accessed 17 January 2015).

Lampietti, J. A., Michaels, S., Magnan, N., McCalla, A. F., Saade, M. and Khouri, N. (2011). A strategic framework for improving food security in Arab countries. *Food Security*, 3(1): 7–22.

Landmatrix (2015). *The Land Matrix Global Observatory*. Available at: www.landmatrix.com (accessed 20 March 2015).

Lankford, B., Bakker, K., Zeitoun, M. and Conway, D. (2013). *Water Security: Principles, Perspectives and Practices*. London: Routledge.

Mahgoub, F. (2014). *Current Status of Agriculture and Future Challenges in Sudan*. Uppsala: Nordiska Afrikan Instituetet.

Mehta, L., G. J. Veldwisch, and Franco, J. (2012). Introduction to the special issue: water grabbing? Focus on the (re)appropriation of finite water resources. *Water Alternatives* 5(2): 193–207.

Mirumachi, N. and Allan, J. (2007). Revisiting transboundary water governance: Power, conflict cooperation and the political economy. *Adaptive and Integrated Water*, pp1–21. Available at: www.newater.uni-osnabrueck.de/caiwa/data/papers session/F3/CAIWA-FullPaper-MirumachiAllan25Oct07submitted2.pdf (accessed 23 May 2015).

Molden, D., Awulachew, S. B., Conniff, K., Rebelo, L.-M., Mohamed, Y., Peden, D., Kinyangi, J., van Breugel, P., Mukherji, A., Cascão, A., Notenbaert, A., Demise, S. S., Neguid, M. A. and el Naggar, G. (2009). Acknowledgements. CPWF Project Report CPWF Project Report, 59.

Mosse, D. and Sivan, M. (2003). *The Rule of Water: Statecraft, Ecology and Collective Action in South India.* Oxford: Oxford University Press.

National Investment Authority (2013). *Directory of Proposed Investment Projects.* Available at: http://sudaninvest.biz/nia/images/documents/pdf/Projects_Book2013.pdf (accessed 12 February 2015).

Newell, P. (2013). *Globalization and the Environment: Capitalism, Ecology and Power.* New York: John Wiley & Sons.

Nile Basin Initiative (2012). *OneRiver One People OneVision*, Entebbe: NBI.

Nile Basin Initiative (2014). *Nile Basin Initiative Corporate Report.* Available at: www.nilebasin.org/index.php/media-center/speeches/cat_view/4publications?orderby=dmdate_publishedandascdesc=DESC (accessed 6 October 2014).

Pattberg, P. and Stripple, J. (2008). Beyond the public and private divide: Remapping transnational climate governance in the 21st century. *International Environmental Agreements: Politics, Law and Economics*, 8(4): 367–388.

Preservenile (2012). Twin Dam in Eastern Sudan: Rumela Dam on Upper Atbara and Burdana Dam on Setit. *Preserve the Middle Nile.* Available at: https://preservethemiddlenile.wordpress.com/2012/04/24/twin-dam-in-eastern-sudan-rumela-dam-on-upper-atbara-and-burdana-dam-on-setit (accessed 4 April 2015).

Ranganathan, R. and Briceño-Garmendia, M. C. (2011). *Africa Infrastructure Country Diagnostic (AICD), Sudan's Infrastructure: A Continental Perspective.* Washington, DC: International Bank for Reconstruction and Development/World Bank.

Scoones, I., Hall, R., Borraas, S. M. Jr., White, B. and Wolford, W. (2013). The politics of evidence : A response to Rulli and D' Odorico. *The Journal of Peasant Studies*, 40(5): 911–912.

Sebastian, A. G. and Warner, J. F. (2013). Geopolitical drivers of foreign investment in African land and water resources. *African Identities* (June 2014): 1–18. Available at: http://dx.doi.org/10.1080/14725843.2013.868669 (accessed 16 September 2014).

Shapland, G. (1997). *Rivers of Discord: International Water Disputes in the Middle East.* London: Hurst.

Siddig, K. H. (2012). *Agricultural Oil and Agriculture in the Post-Separation Sudan Working Paper Series.* Khartoum: University of Khartoum.

Sojamo, S. and Larson, E. A. (2012). Investigating food and agribusiness corporations as global water security, management and governance agents: The case of Nestlé, Bunge and Cargill. *Water Alternatives*, 5(3): 619–635.

Sudan Tribune (2015). Gezira scheme farmers launch election boycott campaign. Available at: www.sudantribune.com/spip.php?article54268 (accessed 30 March 2015).

Taha, F. (2010). The History of Nile Waters in the Sudan. In T. Tvedt (ed.), *The River Nile in the Post-Colonial Age*. London: I.B. Tauris, pp179–216.

Turton, A. (2000). *A Cryptic Hydropolitical History of the Nile Basin for Students of Hydropolitics. Study Guide for Pretoria University*. Pretoria: African Water Issues Research Unit, Study Guide for Pretoria University.

UNDESA (2012). *World Population 2012*. Available at: www.un.org/en/development/desa/population/publications/pdf/trends/WPP2012_Wallchart.pdf (accessed 18 March 2015).

UNEP (2007). *Sudan Post-Conflict Environmental Assessment*. Nairobi, Kenya.

Verhoeven, H. (2011). Climate change, conflict and development in Sudan: Global neo-Malthusian narratives and local power struggles. *Development and Change*, 42(3): 679–707.

Verhoeven, H. (2013). The politics of African energy development: Ethiopia's hydro-agricultural state-building strategy and clashing paradigms of water security. *Philosophical Transactions of the Royal Society of London A: Mathematical, Physical and Engineering Sciences*, 371(2002), pp2–14. Available at: http://rsta.royalsocietypublishing.org/content/371/2002/20120411.abstract (accessed 9 March 2015).

Verhoeven, H. (2014). Is Beijing's non-interference policy history? How Africa is Changing China. *The Washington Quarterly*, 37(2): 55–70. Available at: www.tandfonline.com/doi/abs/10.1080/0163660X.2014.926209 (accessed 21 April 2015).

Verhoeven, H. (2015). *Water, Civilisation and Power in Sudan*. Cambridge: Cambridge University Press.

Vimeo (2014). *Al Kafa'ah Project*. Available at: https://vimeo.com/90346182 (accessed 2 April 2015).

Waterbury, J. (1979). *Hydropolitics of the Nile Valley*. Cambridge: Cambridge University Press.

Wittfogel, K. (1957). *Oriental Despotism: A Comparative Study of Total Power*. New Haven, CT: Yale University Press.

Woertz, E. (2011). Arab food, water, and the big landgrab that wasn't. *The Brown Journal of World Affairs*, xviii(1): 119–132.

Woertz, E. (2013a). Explaining the implementation gap: Money, water, and politics. In Woertz, E. (ed.) *Oil for Food*. Oxford: Oxford University Press, pp227–253.

Woertz, E. (2013b). The governance of Gulf agro-investments. *Globalizations*, 10(1): 87–104. Available at: www.tandfonline.com/doi/abs/10.1080/14747731.2013.760932 (accessed 9 March 2015).

World Bank (2015). Sudan Metadata World Development Indicators. Available at: http://data.worldbank.org/country/sudan (accessed 18 May 2015).

Worster, D. (1983). Water and the flow of power. *Ecologist*, 13(5): 168–174.

Youssef, H., Osman, M. and Roudi-fahimi, F. (2014). *Responding to Rapid Population Growth in Egypt. Population Reference Bureau (PRB)*. Washington, DC: PRB. Available at: www.prb.org.

8 Dams, water and accountability in Uganda

David Ross Olanya

Introduction

This chapter reviews the recent proliferations of hydropower constructions and discusses it from perspectives of state-building and accountability in Uganda. Investing in the energy sector is seen as a priority by the government to meet the transforming *Vision 2040* plan that aims to move Uganda from a peasant economy to a middle-income country through dam construction mainly on the River Nile. These dam constructions can be seen as part of a broader nation-building process that aims to transform the economy in which the energy sector has become a public good that necessitates government attention. The national ambitions to exploit the river through the construction of hydropower are further examined within the contexts of shifting accountabilities between the government, the World Bank and the civil society. This shift includes the new dominant role of China in financing dam construction in Uganda in relation to the World Bank.

Uganda's ambitions to construct several run-off dams as part of a nation-building project has not generated the same hydropolitical contestation from Egypt as, for example, the construction of reservoir dams such as Ethiopia's Grand Ethiopian Renaissance Dam (GERD). The run-off dams under construction in Uganda affect primarily local areas, including water access to people, and the ecology. The displacement and water grabs in the vicinity along the dams in Uganda have received little attention in terms of accountability claims from the public. This raises the question of the land–water–right nexus since the ecology is degraded, and there is inadequate compensation after appropriation and unintended consequences of restricted access to water along the Nile for the poor. In this chapter, the trajectory of hydropower constructions in Uganda is highlighted from perspectives of national-building and accountability. The chapter examines the demand for hydropower constructions in Uganda and discusses the conduct of accountability embedded in the demand for dam construction in Uganda from World War II onwards.

Dams in Uganda

Dams are constructed for different purposes: to irrigate agriculture, to generate electricity, to control floods and to prevent siltation. Dam constructions alter and divert river flows and often affect existing rights and access to water, hence, they significantly impact upon people's livelihoods and the environment (WCD, 2000). In Uganda, dam constructions have mainly been driven by increased economic demand for electricity for the enhancement of the industrial sector, where dam constructions are expected to transform the economy from a peasant economy to a middle-income economy by 2040. The 2040 Vision, which was approved in 2007, is a national development policy that intends to transform Uganda from a peasant society to a modern and prosperous country within 30 years. It was launched in 2013 and addresses key bottlenecks that have constrained the country's socio-economic development since independence. As a nation-building project, agriculture, infrastructural development, water resources and industrialization are among the top priorities within the 2040 Vision. The vision proposes to improve access to electricity from the current 11 per cent to 80 per cent in 2040. The transformative *Vision 2040* emphasizes Northern Uganda as a future industrial hub for the country in which hydropower is seen as central for the vision (National Planning Authority, 2013).

Similar to other Nile riparian states where dam constructions have been constructed, these dams can also be conceived as instrumental for the development of Ugandan nationalism and nation-building. Previous dams in Uganda have been projects with storage reservoir. The recent dam constructions in Uganda are mainly run-off: dams where little or no water storage is needed to generate electricity (WCD, 2000: 11). In Uganda, there is increasing interest by the government to meet energy demand through the construction of large-scale infrastructure projects along the River Nile. Dam proliferation is driven mainly for industrial and agricultural projects to boost the country's economic growth.

In addition to the hydropower projects already under construction, Uganda has several planned hydropower projects, including Kiba (300 megawatts [MW]), Uhuru (350MW), Oryang (392MW), Murchison Falls (700MW) and Ayago Falls (700MW) (Table 8.1). These projects are all located in Northern Uganda and in the Bunyoro region. Within the geospatial vicinity of these projects are large-scale acquisitions of land. Grain milling factories are being built in Nwoya as part of large-scale farming projects. A sugar factory (Amuru Sugar Works Limited), although delayed, will be built in Amuru district and another one is coming in Atiak sub-county. All these agro-processing businesses demand electricity supply.[1] It is in this context that the government plans to construct the large hydropower station located in Murchison Falls National Park, Nwoya district, in Northern Uganda. The construction is expected to commence after 2018 when oil revenues start flowing into the economy.

Table 8.1 Hydropower in Uganda*

Name of project	*Construction phase*	*Type of construction*	*Power capacity*
Nalubaale	Completed 1954	Reservoir, Lake Vitoria	180MW
Kiira	Completed 2000	Reservoir, Lake Victoria	200MW
Bujagali	Completed 2012	Run off river	250MW
Karuma	Under construction to be completed in 2018	Run off river	600MW
Isimba	Under construction to be completed in 2018	Run off river	183MW
Ayago	Under construction to be completed in 2023	Run off river	600–840MW
Murchison Falls	Anticipated project start 2018	Run off river	700MW
Oryang	Planning stage	NA	392MW
Uhuru	Planning stage	NA	350MW
Kiba	Planning stage	NA	300MW

Note

* Besides the hydropower presented in the table above, there are mini-hydro stations: Kikagati (16MW), Mitaano (2.9MW), Lubilia (5.4MW), Nyagak III (4.5MW), Siti (21.5MW), Waki (5.4MW), Rwimi (10.54MW), Ndugutu (0.5MW), Nkusi (4.8MW), Nyamwamba (9.2MW), Neengo Bridge (6.5MW), Esia (0.11MW), Muziizi (44.5MW peak power), Nshungyezi (38MW), Ntono (2.5MW), Sindoro (5.6MW), Aswa-Agago (88MW), Kakaka (7.2MW), Muvumba (4.5MW), Muyumbe (3.2MW), Kyambura (8.3MW), Maziba (1.0MW), Solar (20MW) and additional bagasse from Kinyayara (25MW). The total generation will amount to 1,974MW compared to 60MW in 1986.

Hydropolitical contestation often surfaces in contexts where reservoir dams are constructed with the ability of securing water storage during seasons of water scarcity. Egypt has so far not contested the constructions of run-off dams along the Nile in Uganda since these dams are not considered to impinge negatively on the water flow of the Nile. Thus, in Uganda, run-off river dams do not pose the same threat to Egypt's water security as the reservoir dams currently being put in place along the Blue Nile.

Dams and accountability

Accountability is considered as a concept depicting how governments, the World Bank and the civil society ensure that dam construction respects local rights to water resources and other ecological parameters (see Oestigaard, 2015). This section traces the shifts in accountability between the nation state (internal accountability), the World Bank and civil society under the Environmental Impact Assessment (EIA) mechanism with regards to dam constructions in Uganda. Egypt and other downstream countries have enhanced their water security by large dam reservoirs. Historically, increasing demand for water security and the need for more hydropower have initiated dam construction, and this has been common among other riparians suffering from electricity shortages despite the potential availability of Nile water in their territories – such as Ethiopia and the newly independent state of South Sudan.

Pax Britannica established a Protectorate over Buganda in 1894 and its authority was asserted over other areas in 1896. This was reinforced by the arrival of Sir Harry Johnston in 1899 and, in 1900, the Buganda Agreement. Johnston submitted a prospective report about his good impression of the Protectorate, including climate, people and the resources – in addition to mapping the altitudes as well as navigational rivers and density of the population.[2] The report enhanced the strategic interest of the Nile in Uganda. It was also in Egypt's interest since about two-thirds of the summer Nile water reaching her territory came from Equatorial lakes in the British Protectorate. Egypt's interest in Uganda's water was therefore inevitable from the beginning of the twentieth century. Furthermore, it was in Egypt's interest that the Land and Survey Department was established to promote hydrological science on the River Nile in Uganda until 1948 when a separate Hydrological Department was set up and renamed the Water Department in 1956 (Morris, 1978).

In Uganda, the construction of the Owen Falls Dam, inaugurated in 1954, was mainly driven by the colonial government that aimed to boast the economy of Uganda as a self-reliance country after World War II (Morris, 1973). The Owen Falls Dam was constructed during a time when commodity exports received relative high values at world markets, enabling other development projects under the Colonial Development Act of 1940. The construction of the Hydro-Electric station at Jinja was expected to provide cheap and abundant electricity to the newly established industries, including the lime deposit works at Tororo and the Kilembe cooper mines (Morris, 1973: 27–28). Historically, the quest for more power generation was thus driven by economic factors, but such economic motives are often presented as political to rally support behind the projects.

Economic prosperity often tends to reinforce a sign of political contribution to national-building processes where hydropower constructions along the Nile also are intertwined in accountability struggles between riparian states. In this case, when Uganda constructed the Owen Falls Dam, exchanges of notes between Great Britain and the government of Egypt took place on 30 and 31 May 1949, as part of the 1929 Agreement. During the construction of the Owen Falls Dam, now renamed Nalubale, internal accountability was between the colonial master, Great Britain, Uganda and the government of Egypt in securing its water security. The accountability between the riparian states is reflected in the terms of conditions of the Owen Falls Dam:

1. primarily serving the purpose for hydro-electricity and control of the Nile water;
2. approval of plans and details by Egyptian ministry of works and Ugandan authorities, upon which the contract was initiated by Great Britain;

3 both Great Britain and Egypt examined the contracts jointly and approved, and then notified the Government of Uganda;
4 both governments agreed on the construction of the dam, with responsibility given to Uganda Electricity Board, but a resident Egyptian engineer was stationed at the site to guarantee the interest of Egypt, and after completion, administration and maintenance were the work of Uganda Electricity Board; regulation of water discharges was conducted on the instruction of an Egyptian resident engineer;
5 the Uganda Electricity Board could consider taking action that would not jeopardize the interests as per the 1929 Agreement, and differences would be a matter of discussion and settlement in the spirit of friendly cooperation through arbitration.[3]

The construction of the Owen Falls Dam drew attention to Lake Victoria, the third largest lake in the world (if the Caspian Sea is included) in addition to the other inland lakes along the River Nile, such as Kyoga and Albert. Lake Victoria covers appropriately 68,500km^2 forming part of four countries – Uganda, Kenya, Rwanda and the Republic of Tanzania (Kite, 1982: 99). These large, inland lakes have potential interest for the Egyptian government, with Lake Victoria as the centre of attention where several hydrologic investigations took place from 1967 onwards, which were also expanded to Upper Lakes Kyoga and Albert, and rivers (WMO, 1974, 1977). The riparian members were interested in investigating monthly lake levels and outflows to understand alternative patterns of regulation of Lake Victoria. The run-off and outflow of the lake was measured as well as precipitation and evaporation over the lake. Findings were considered promising in terms of the relative magnitude of the water balance – both rainfall and evaporation over the lake; while run-off and lake outflow were much smaller (Kite, 1982: 100). It is worth noting that one of the major reasons for the formation of the hydro-metrological project was due to the rise in Lake Victoria's levels during the early 1960s – in which the level increased from 1,134 metres above sea level (masl) in January 1960 to 1,136msal in June 1964 (Kite, 1982: 102). The rise in the level prompted the need to carry out hydrologic surveys where Egypt provided the logistical and scientific assistance; fortunately, it was found that the rise in the level of Lake Victoria was not man-made (Kite, 1981).

Dams and water grabs

While substantial land grabbing takes place in Uganda, including 100,000ha of Bukaleeba Central Reserves, 20,000ha of Luwunga Forest Reserves, 40,000ha of land by Wilmar International, 10,000ha of land by BIDCO company in Buvuma Island and 20,000ha of land in Amuru district (Olanya, 2014: 78), these are to a large extent located in rain-fed areas and hence different from large irrigation fields further north in the Nile Basin. What takes place in Uganda has thus different political implications for Egypt's water security and for the Nile Basin as a whole.

During the past decades, literature on the effects of dam construction have focused on physical displacement and dispossession of peoples from their lands. The dispossession arises when the government only justifies dam construction in terms of technical planning and nation-building to the extent that it connotes displacement, dislocation, eviction, exclusion and involuntary resettlement as a matter of internal accountability. Displacement is, however, often used to signify involuntary physical removal of the people from their historical or existing homes or areas due to the decision to develop the areas by the government and its partners. In addition, there could be a considerable loss of access to or restriction on livelihood opportunities or future income related to changing environments. In this chapter, I see the issue of displacement, particularly in situations where the public stake is high on producing energy, as constituting a special form of water grab in which the government avoids being accountable to those who are directly or indirectly affected by dam construction. Such water grabs often emerge in situations when the implementation of hydropower projects tend to focus more on technical aspects rather than on socio-economic impacts. It is from this perspective that dam constructions generate winners and losers, and, hence, when what has happened to those losing are not adequately addressed.

While this chapter bases itself within the national frame, emphasizing the importance of dams in sustaining growth progress, consequential efforts such as involuntary displacement and rights to land and water and other ecological changes are often shelved in favour of national interests and agendas. In the Ugandan context, accountability tends to concentrate on rationalistic arguments in technical and engineering components rather than the socio-economic impacts. While the trend in dam construction has concentrated on development-induced displacement, this conception is rather limiting in the context of the growing trend focusing on accountability and rights and access to resources of the communities in the project areas. This contribution shows how dam construction has shifted the debate on water grabs and on accountability by the government to other resource users. In the past decades, the government faced little accountability challenges from the affected communities, funders and internal environmental groups. Rather, the government has paid more attention to state-building than to the needs expressed by local communities being displaced during dam construction. The development of dams is given national pride and priority in Uganda, but little attention focuses on compensating prior water users in addition to having alternative clear policies in place.

Hydropower as a state-building process and power consolidation

The completion of the Owen Falls Dam in the 1950s was demand driven in order to promote economic development. It was seen as a national symbol of championing economic transformation in Uganda. Construction also contributed to reinforce a moral liberal reasoning that the state should provide the public goods necessary for the economic transformation of the country. In a similar vein as in the past, today's transformative *Vision 2040* seeks to transform the country from a peasant to middle-income country and is framed in a nationalistic state-building discourse. The 2040 transformative vision has a high stake among the political and policy elites in the energy sector. The 2040 Vision particularly points to hydropower generation along the River Nile as a cornerstone for rebuilding the national economy. Hydropower generation has become one of the 'imagined' public goods to be provided in the next decades. This reminds us of what Billington and Jackson (2006) called 'monument to democracy' in the United States when dam construction was seen as a monument to democracy in the welfare state. Interest groups mobilized and favoured dam construction to control floods, enable irrigation and facilitate navigation under New Deal philosophy in the United States. Dam construction had progressive impacts in regional economies and planning by which the state elites and economic actors became very influential in determining the outcomes. Dam construction under the New Deal created employment and was considered an accomplishment of technological advancement, but how the projects dislocated local communities was a secondary issue to be handled technically by the visioning state.

However, in the recent past, there has been a shift in accountability from the government to the World Bank and civil society regarding those affected by mega-projects in the global South, especially when projects were funded by the International Finance Corporation under the World Bank. This change in accountability emerged in 1985, when the World Bank had to accommodate two main outside groups – the United State Congress and the Northern non-governmental organizations (NGOs), in addition to the existing internal accountability. Borrowing countries in the global South had to be accountable to NGOs. This is what Fox and Brown have called 'non-profit civil society organizations that do grassroots support and advocacy' (2000a: 21). In this situation, the process of accountability at the World Bank shifted away from *internal accountability* by a borrowing country towards embracing accountability (*external accountability*) that placed demands on various international advocacy groups, NGOs as well as accountability demands from the US government.

Lending governments now had to be accountable to the World Bank and their own constituencies, and also, in this case, complaints by advocacy NGOs such as Friends of the Earth and International Rivers Networks that were very critical of the World Bank and raised awareness of environment and social issues. NGOs were working in collaboration with the World Commission on Dams, the Independent Inspection Panel and even the World Bank itself. Financers of mega-projects had to accommodate complaints from both Northern and Southern NGOs (Fox and Brown, 2000b) who engaged the World Bank and the borrowing country to respect and commit themselves in the forms of an EIA and an Independent Inspection Panel (Linaweaver, 2009: 285).

It is worth noting that it was the World Bank itself that introduced the EIA policy in 1989 as a response to environmental groups in Washington in order to respond to the growing concern for environmental issues. In addition, US Congress had also passed legislation that required multilateral banks to 'encourage borrower countries to fully inform local communities and appropriate NGOs of project planning, to allow them to participate' (Linaweaver, 2009: 286). The World Bank procedures (Operational Procedure 4.01) also demanded that borrowers 'consult project-affected groups and local non-governmental organizations (NGOs) about the project's environmental aspects and take their views into account' (World Bank, 1999; Linaweaver, 2009: 286). Participation was defined as a process through which stakeholders influence and share control over development (Nelson and Wright, 1995: 5). The following section outlines how the shift from internal accountability to external accountability has affected Uganda as a borrowing country in the implementation of the Bujagali hydropower project (see also Oestigaard, 2015).

Bujagali Falls hydropower project

The Bujagali hydropower project was proposed in 1994 along the River Nile, 8km downstream from the Owen Falls extension and was completed in 2000. Bujagali hydropower was projected at US$550 million and it was expected to produce 250MW of electricity. The Bujagali hydropower project is located on a former popular tourist site. In addition, the Bujagali hydropower project had displaced 1,288 households in Busoga on the East Bank of the Nile and on the west side of the river in Buganda. Based on the principle of *internal accountability*, the Bujagali project was awarded to US-based AES Corporation. The subsequent companies were also engaged in other parts as a long-term strategy to achieve the construction of 250MW of electricity with financing from the International Finance Corporation on 18 December 2001 (Linaweaver, 2009: 288).

The shift in accountability requirements by the World Bank delayed the construction of the Bujagali hydropower project. The World Bank, the Ugandan government and the US-based AES Corporation conducted 1,300 consultations with interested groups, local residents and cultural leaders during a period of seven years (Chestnut, 2001). The Bujagali project received extensive criticisms from international NGOs, including International River Networks and Friends of the Earth in alliance with the National Association of Professional Environmentalists, Save Bujagali Crusade and the Busoga on the East Bank of the Nile River. In ensuring that consultation took place, the involvement of the International Rivers Networks and the National Association of Professional Environmentalists shifted in-country accountability from the government of Uganda to the World Bank (Linaweaver, 2009: 288).

The *intra-country accountability* was also questioned by the Busoga kingdom due to the destruction of significant heritage sites in Busoga kingdom. Moreover, when Ripon Falls was submerged in the 1950s after the construction of the Owen Falls Dam (now Nalubale), it led to the displacement of people who were not sufficiently compensated. Against this background, the US-based company AES Corporation had to develop a *Cultural Property Management Plan* taking into consideration people's cultural and traditional religious beliefs associated with the destruction of the falls. The plan was basically designed on how to move 'spirits' from the construction site (Lacey, 2001). The issue of transfer of accountability from the government to the World Bank became more prominent when the International Rivers Networks began to lobby the World Bank in 1999 on issues regarding the lack of transparency of the power purchase agreements and the potential excessive electricity tariffs Uganda could experience when the Bujagali project was completed. Arising from this external pressure, the government and the World Bank were made accountable to the interests of International Rivers Networks to such an extent that it reduced the government's own accountability to borrow money from the World Bank. The engagement of actors from the civil society with the World Bank in ensuring participation by the local community has reduced the sovereignty of borrowing countries. The presence of an excessive accountability by the World Bank also determined the process of conducting an EIA that otherwise would only have been left to the borrowing countries. In this regard, the EIA processes should have had a local regulatory role of the borrowing government rather than being the work of the World Bank in order to avoid a development hold-up (Linaweaver, 2009: 284). The strict guidelines from the World Bank on the borrowing government during the implementation of the Bujagali hydropower project were necessary requirements for financing the energy sector in Uganda. However, the excessive World Bank accountability requirements caused frustration in Uganda, and the government of Uganda blamed the delayed Bujagali project as one of the primary factors contributing to the country's persistent energy crisis. When the government failed to meet the accountability requirements from the World Bank, the energy instability tightened its grip on Uganda's economy. The government aimed to complete the Bujagali project and argued that 97 per cent of Ugandans had no access to electricity amidst growing demands for electricity. In this regard, hydropower contributes 83 per cent of the Uganda's electricity supply, and Uganda currently imports electricity from Kenya and Rwanda.

The delayed Bujagali hydropower project had two important lessons for the government: the high electricity cost for industrialization and losses in transmission; and the low electricity cost for the industrial sector, necessary to boost economic progress. Although the government maintains its position that the industrial and manufacturing sectors need cheaper electricity, the electricity charge for these sectors is nevertheless more than US$5 per unit. In promoting competitive economy on the global market, factories need lower energy prices, creating more jobs and adding value to domestic raw materials, and, in the end, generating more money for the country. Unfortunately, the energy from Bujagali became more costly than the anticipated US$4.9 per megawatt hours (MWh) and it went up to 10.1MWh. This rise in the electricity tariffs was a consequence of the high capital investments by developers whom the government had borrowed money to complete the project in addition to high costs of construction materials (steel, oil) while building the dam (State of the Nation Address, 2015). Still, the government maintains its stand on resisting distraction from the path towards development progress. By focusing on development and wealth creation, the government prioritizes electricity generation through expanding its distribution lines from 1,427km in 2006 to 1,627km in 2015. The medium voltage lines (33KV and 11KV) totalled 15,178km in 2015 compared to 6,245km in 2006. Consumers' connectivity to lines has increased from 8,448km to 18,000km in 2015, and all 112 districts will be connected by 2017. The government has also capitalized politically on hydropower generation by framing electoral politics as part of power generation, as illustrated in Figure 8.1.

The Bujagali hydro-electric project as a multi-million project was designed to add 250MW of electricity to the national grid. The US$860 million hydropower project was commissioned by President Yoweri Museveni as part of strategic infrastructural development in Uganda.[4] Investing in energy has top priority for the current regime and most of the projects are constructed along the River Nile. Since 2006, Uganda has been experiencing an energy crisis when the level of Lake Victoria reduces significantly. The government adopted a short-term strategy by engaging private developers in thermal power plants to the extent that the level of subsidy became unsustainable, amounting to 1.5 trillion Ugandan shillings by 2015. It is against this background that the government abandoned its short-term strategy and embarked on a long-term nation-building process through the construction of run-off dams. It is believed that the projects will reduce the incidence of low power supply to the national grid, which has resulted in high costs for those doing business in Uganda.

Simplifying accountability: China in Uganda's energy sector

Karuma hydropower project

In December 2013, Uganda embarked on yet another hydropower construction, the so-called Karuma hydropower project. The Karuma hydropower project is expected to produce 600MW of electricity by 2018 and is being constructed by SinoHydro Corporation. The project is expected to cost between US$1.6 billion and US$2.2 billion and for this project the government of Uganda has sought alternative sources of financing. About 85 per cent of the funding comes from the Export-Import Bank of China and the Ugandan government will cover the rest, including US$253 million to kick-start the project. The EXIM Bank of China is providing the funding through two-tier loan schemes. The first funding scheme is estimated to be US$789 million at an interest rate of 2 per cent per annum with a 20-year maturity and a five-year grace period. The second loan funding scheme is estimated at US$646 million. The project has so far experienced slow progress, arising from logistical challenges between the funder EXIM Bank of China and the government

Figure 8.1 Uganda's industrial development is largely anchored on hydropower generation

of Uganda.[5] Although the project started slowly, it has only been subjected to internal accountability challenges, and the Ugandan Parliament approved the loans in March 2015. By engaging with the Chinese Development Bank, the government avoided going through the excessive and external accountability mechanisms under the World Bank and the actors from the civil society that had delayed the construction of the Bujagali hydropower project. Rather than being exposed to multiple accountability claims that dominate the World Bank's funding in the global South, accountability became an internal matter of the government in relation to the funder, the EXIM Bank of China. According to the government, the tariffs from the Karuma hydropower project will be US$5 per MWh. The government is committed to achieving affordable and cheap electricity supply for both domestic and industrial purposes. Still, Uganda sells electricity to factories at US$11 per unit. Moreover, losses in transmission under the Uganda Electricity Board were at 34 per cent, although President Museveni claims the losses have been reduced to 14.7 per cent, still an unsatisfactory number.[6]

The Karuma hydropower project is a run-off-river type hydro-generation project depending on underground technology to generate 600MW of electricity, and consists of six turbines, each of them designated with an independent water tunnel. In a similar vein as with Bujagali, the ecological consequences of the project have been downplayed in the public policy debate. The Karuma project is located within Murchison Falls National Park, accommodating a rich biodiversity of fauna and flora. The government and the funder (EXIM Bank of China) have, on the other hand, given a lot of attention in the planning process to technical and construction details rather than to the socio-economic impacts of the hydropower project, including dams as water grabs.

There are also other parallels to Bujagali. When the Karuma hydropower project was first proposed in 1995, the World Bank prioritized the Bujagali project instead. Similarly, when the Norwegian company Norpak was granted the contract to build the 250MW dam at Karuma Falls, the company had to pull out in 2009 because of the alleged corruption scandal. The International Rivers Networks argued that the Karuma hydropower project affected local communities, which had to bring their concerns to the government over unfair compensation from the devaluation process of land and property. The Karuma hydropower project affected four villages: Karuma, Awoo, Nora and Akurudia. The Electricity Disputes Tribunal was petitioned because of meagre compensation. In most cases, prioritization was given to engineering solutions rather than mitigation issues arising from the social economic impacts of displacement and resettlement.[7]

Isimba hydropower project

Isimba hydropower project (183MW) is being constructed 40km downstream from the newly constructed Bujagali hydropower project. Isimba is also funded by the Chinese Export and Import Bank (EXIM) and China International

Water and Electric Corporation at a cost of US$500 million. The construction of the Isimba Dam hydropower project in Kayunga district and its 132KV transmission line will destroy 18km of the river used for touristic white-water rafting on the River Nile. Those working in tourism and the rafting industry argue that the Isimba Dam destroys a popular sport site on the River Nile attracting huge numbers of tourists. At least 18km of the river will, however, remain unaffected by the construction, including the stretch between Bujagali and Kalagala Falls. Still, the construction of the Isimba Dam will also cause flooding in five other rapids such as Weleba, Malalu, Nile Special, Kulu Shaker and Hair of the Dog. As a result, Weleba and Malalu rapids will have less value for rafting downstream, and the Isimba Dam, as Bujagali, is expected to have huge economic consequences for the multi-million dollar rafting industry.[8]

Prospecting hydropower projects along River Nile: petrodollar financing

The transformative *Vision 2040* has a high stake among political and policy elites in Uganda. The vision aims to move Uganda from a peasant to an industrialized modern economy by emphasizing hydropower. The ongoing hydropower projects (run-off dams) along the River Nile in Uganda – for example, the Karuma hydropower project (600MW) and the Isimba hydropower project (183MW), being financed by China and the government having committed up to US$338.4 million of the energy fund – have changed the priorities of the transformative state of Uganda in a similar ways that have taken place in other riparian countries along the Nile. The GERD is being funded by the Ethiopian government through selling dam bonds directly to citizens both at home and abroad and to private companies.[9] In Uganda, in addition to the funding from China, there are a number of prospecting hydropower projects along the Nile such as the Ayago hydropower project (840MW) and Murchison Falls (700MW), which are planned to be financed by future oil revenues. When the oil revenues start flowing in 2018, priority will be given to the additional electricity-generating capacities at Ayago (840MW), Oryang (392MW), Uhuru (350MW), Kiba (300MW), Albatros Thermal Plant (50MW) and Murchison Falls (700MW). According to the government, a total of 4,356MW installed capacity will be realized when the oil revenue starts flowing and the country will be in a position to pay off the expensive loans borrowed for the construction of the Bujagali hydropower project.

These planned dams are all run-off dams and very different from those previously being built along the River Nile. Earlier dams on the Nile have reservoir lakes, such as the Owen Falls Dam, the Aswan High Dam and even the newly constructed GERD. Run-off dams minimize the environmental impacts and the displacement of local communities. Still, run-off dams tend to be controversial, particularly in the areas where the projects are constructed. The Karuma hydropower project is, for example, located within Murchison

National Park and affects the ecology and the environment, especially the flora and fauna within the conservation area. Unlike the Bujagali project that had multiple accountability mechanisms – to the World Bank's EIA and actors from the civil society – the Karuma hydropower project has received comparably little attention from civil society. In the case of the Karuma hydropower project, the political stakes are high and the government strongly emphasizes production of electricity for Ugandans. Moreover, the effects of the Karuma hydropower project on nature conservation are hardly presented in EIAs. In most cases, environmental impact studies for reservoir dams are used by opponents in other riparian countries, like the GERD is in Egypt and Sudan. When Uganda concentrates on run-off dams, those contesting the constructions are mainly people living in the nearby areas directly affected, as well as associated NGOs.

Conclusions

This contribution has highlighted how the proliferation of hydropower projects in Uganda is framed within a broader nation-building process that aims to transform the economy from a peasant economy to a middle economy by 2040. While dam construction inevitably involves the displacement of people and internal activism as well as opposition from foreign NGOs, in the overall Nile Basin geopolitics, the Ugandan hydro-electrical dams have received less international attention due to shifts in accountability requirements. When Uganda turned to China instead of the World Bank for funding, the Ugandan government received less international exposure from advocacy groups. The shift in financing and the collaboration with Chinese corporations have reduced the participation of environmentalists and decreased the ability of local communities in the decision-making process. Moreover, the run-off dams constructed along the Nile in Uganda to a large extent mostly impinge negatively on local water users and ecologies near the project, and may not harness the interests of downstream users to the same extent as dams that require large water reservoirs.

Arguably, criticism targeting World Bank's financing of dam construction transformed the World Bank's policy in 1982 to adopt an internal directive, including the interests of indigenous peoples, especially regarding resettlement and environmental assessments (Linaweaver, 2009). It further established an appeal mechanism in 1993 that allow communities adversely affected by the Bank's funding of projects to file complaints regarding violation of policies, procedures and loans arrangements (WCD, 2000: 19). However, this trend has changed regarding the contemporary dam constructions in Uganda given the shift in dam financing from World Bank sources to public sources and China, in addition to the decline in donor funding and the increased costs of dams (WCD, 2000: 20). The reduction in external funding reduces the extent of external accountability by the host government as internal accountability takes shape.

Notes

1 State of the Nation Address, His Excellency Yoweri Kaguta Museveni, President of the Republic of Uganda at the Uganda International Conference Centre, Serena, Kampala, 4 June 2015.
2 Dispatch of 27 August 1901, F.O. 2/40; see also Morris (1973: 16).
3 No. 3122. Exchange of notes constituting an agreement between the government of the United Kingdom of Great Britain and Northern Ireland and the government of Egypt regarding the construction of the Owen Falls Dam, Uganda. Cairo, 30 and 31 May 1949.
4 *The New Vision*, 8 October 2012, Museveni Commissions Bujagali Dam.
5 See http://china.aiddata.org/projects/30919 (accessed July 2015).
6 State of the Nation Address, His Excellency Yoweri Kaguta Museveni, President of the Republic of Uganda at the Uganda International Conference Centre, Serena, Kampala, 4 June 2015.
7 See www.internationalrivers.org/resources/uganda-s-dams-leave-affected-people-behind-8083 (accessed July 2015).
8 *Daily Monitor*, Isimba dam to swallow rafting site, 29 June 2015.
9 See www.internationalrivers.org/resources/the-grand-ethiopian-renaissance-dam-fact-sheet-821 (accessed July 2015).

References

Billington, D. P. and Jackson, D. C. (2006). *Big Dams of the New Deal Era: A Confluence of Engineering and Politics*. Norman: University of Oklahoma Press.

Chestnut, J. (2001). Good Communication Help Social Impact Studies at Bugali. *Hydropower and Dam*, 6.

Fox, J. and Brown, D. (eds) (2000a). *The Struggle for Accountability*. Cambridge, MA: MIT Press.

Fox, J. and Brown, D. (2000b). Accountability within Transnational Coalitions. In Fox, J. and Brown, D. (eds). *The Struggle for Accountability*. Cambridge, MA: MIT Press.

Kite, G. W. (1981). Recent Changes in Level of Lake Victoria. *Hydrological Sciences Bulletin*, 26(3), 233–243.

Kite, G. W. (1982). Analysis of Lake Victoria Levels. *Hydrological Sciences Journal*, 27(2): 99–110.

Lacey, M. (2001). Traditional Spirits Block a $550 Million Dam Plan in Uganda. *New York Times*, 13 September.

Linaweaver, S. (2009). Catching the Boomerang: EM, the World Bank and Excess Accountability – A Case Study of the Bujagali Falls Hydropower Project Uganda. *International Journal of Sustainable Development and World Ecology*, 10(4): 283–301.

Morris, H. F. (1973). *Introduction to the Microfilm Uganda Land Reports from the Commonwealth Office*. Wakefield: EP Microform.

Morris, H. F. (1978). *Introduction*. Wakefield: EP Microform.

Nelson, N. and Wright, S. (1995). Participation and power. In Nelson, N. and Wright, S. (eds). *Power and Participatory Development: Theory and Practice*. London: Intermediate Technology Publications, pp. 1–12.

National Planning Authority (2013). *Uganda Vision 2040*. Available at www.npa.ug/vision2040/vision-targets.htm (accessed January 2015).

Oestigaard, T. (2015). *Dammed Divinities: The Water Powers at Bujagali Falls, Uganda*. Current African Issues No 62. Uppsala: Nordic Africa Institute.

Olanya, D. R. (2014). Asian Capitalism, Primitive Accumulation and the New Enclosures in Uganda. *African Identities*, 12(1): 76–93.

State of the Nation Address (2015). His Excellency Yoweri Kaguta Museveni, President of the Republic of Uganda at the Uganda International Conference Centre, Serena, Kampala, 4 June.

WCD (World Commission on Dams) (2000). *Dams and Development: A New Framework for Decision Making. The Report of the World Commission on Dams*, November 2000. London and Sterling: Earthscan.

WMO (1974). *Hydrometrological Survey of the Catchments Lakes Victoria, Kyoga, and Albert.* 3 vols. WMO, CP no 5, CH-1211.

WMO (1977). *Hydrologic Model of the Upper Nile Basin.* Report in 7 volumes prepared by Snowy Mountains Engineering Corporation for WMO, Geneva.

World Bank (1999). *Operational Policy 4.01 Environmental Assessment, in the World Bank Operational Manuel.* Available at http://siteresources.worldbank.org/INTFORESTS/Resources/OP401.pdf (accessed July 2015).

9 'Lease the land, but use the water'

The case of Gambella, Ethiopia

Wondwosen Michago Seide

Introduction

Gambella is one of the regional states in Ethiopia where huge tracts of land have been leased out to domestic and international investors since the global food, fuel and financial crisis of 2007/2008. The Gambella region is also one of the most marginalized regions in Ethiopia. It has been alienated from the central politics and socio-economic activities of the state. In the past, the Gambella region was mainly known in relation to ethnic conflicts between the Nuer and Anywaa ethnic groups and the highlanders and in relation to the civil wars in Sudan and later in South Sudan. At the moment, large-scale agricultural investment commonly labelled as 'land grabbing', depending on which perspective one takes, has brought the Gambella region to the centre of national and international attention. Several reports, newspaper articles and policy papers (Dessalegn, 2011; Oakland Institute, 2011; *The Guardian*, 2011, 2015; *Time Magazine*, 2011; GRAIN, 2012; Mittal, 2013; Nalepa, 2013) often refer to land and water grabbing activities in Gambella Regional State (GRS). Still, an overview of different 'land investment deal contracts' in Ethiopia reveals that water is either vaguely mentioned or is implicitly given out to leasers together with the land.

According to the International Food Policy Research Institute, 'increased biofuel demand in 2000–7 is estimated to have contributed to 30 per cent of weighted average increase of cereal prices' (von Braun *et al.*, 2008, cited in Cotula *et al.*, 2009: 53). Agribusiness, investment funds and government agencies in general have been investing in mainly developing countries, including Ethiopia (Cotula *et al.*, 2009; Cotula, 2011). These investments have been reported to have both positive and negative impacts. The problem associated with commercial farming in Gambella, as Oakland Institute (2011, 2013) GRAIN (2008, 2012), and Human Rights Watch (2012) argue, is that it creates 'property insecurity' resulting in scarcity of water and grazing land. The positive side, as the World Bank argues, is that it can create opportunities for recipient countries and local people, by bringing capital, know-how, jobs, and market access and infrastructure development. From this perspective, large-scale land investment may reduce rural poverty and deliver social and environmental benefits (World Bank, 2010; Deininger, 2011, cited in Mehta *et al.*, 2012).

Land and water are intrinsically linked. There is no commercial farm without reliable water resources. This is why land grab is also a water grab. The most often quoted water grab definition was the one given by the Global Water Grab: 'Water grabbing refers to a situation where powerful actors are able to take control of or reallocate to their own benefit, water resources at the expense of previous (un)registered local users or the ecosystems on which those users' livelihoods are based' (Franco *et al.*, 2014: 3). Capturing water is not only controlling the water, but also a reflection of 'the power to decide who, when and how to use the water' (Franco *et al.*, 2013: 153). It involves the unequal power relations between the user, provider and the local peoples as to how, when and how much of the water to use (Franco *et al.*, 2013: 153). Throughout human history, competition over water resources has always existed. The contemporary wave of water grabs is occurring because of the land grabs, which, in turn, may result in water resource appropriation and enclosures and the conversion of water from 'public/common goods' to private/commoditized goods. This has a 'potential drastic impact on the current and future uses and benefits of water, rights as well as changes in tenure relations' (Franco *et al.*, 2013: 3).

Numerous research (Smaller and Mann, 2009; Woodhouse and Ganho, 2011; Skinner and Cotula, 2011, cited in Mehta *et al.*, 2012: 3) shows that 'land grabbing may be motivated by the desire to capture water resources'. Despite the large scholarly debate and media attention focusing on 'land grabbing' or 'land investment',[1] there is limited research on 'water grabbing' or water use in these deals. The existing studies on 'water grabbing/use' mainly focus on the use of upstream water for land investment and its impact at the intra-state level (see Smaller and Mann, 2009; Jägerskog *et al.*, 2012; Woodhouse and Ganho, 2011, cited in Jägerskog *et al.*, 2012). In Gambella, there is even more limited empirical research done on the subject. There is almost no study on the impact of land acquisition on transboundary water resources and its implication on the 'hydropolitics' of the Nile River Basin (see Chapter 2 in this volume). This study aims to explore the hydropolitics and local practices by analysing land and water acquisition in the GRS and along the Baro-Akobo River, which is a tributary of the Nile River.

This study is based on key informant interviews with policy-makers and government officials at both federal and regional government levels, including the Ministry of Agriculture (MoA), Ministry of Water, Irrigation and Electricity in Addis Ababa, and bureaus working with agriculture, environment, investment, trade and industry in the Gambella region. The study also draws on data gathered from investors and, more importantly, interviews and focus group discussions with local community members. The author undertook two field visits to the Gambella region, including visits to the Nuer and Anywaa zones in 2012 and 2013, and also made personal observation to a number of commercial farms owned by both domestic and international investors.

The chapter is organized as follows. The section that follows these introductory remarks gives a brief description of the GRS. Then the chapter proceeds to discuss the land tenure system in the region and large-scale land acquisition in Ethiopia in general and the GRS in particular. The last main section examines water use and grab in the GRS.

Study area

The comprised area of Ethiopia is estimated at about 1.13 million km^2 of which 99.3 per cent constitutes land area and about 0.7 per cent water bodies (MoWR, 2002, cited in Awulachew *et al.*, 2007). The current population of Ethiopia is about 96 million, of which more than 84 per cent live in rural areas (EEA, 2015). Ethiopia is a federal democratic republic consisting of nine national regional states and two administrative cities.[2] The GRS is one of the regional states that have two administrative zones and eight *woredas* (districts). The GRS is located in the south-western Ethiopian lowlands, bordering South Sudan, and covers a total area of about 25,294km^2. Ninety per cent of the population of Gambella is rural and most are subsistence farmers and pastoralists. The ethnic groups inhabiting the region include the Nuer, Anywaa, Majangir, Opio and Komo as well as those from the Ethiopian highlands, who settled in the region from the 1970s and 1980s. Based on the 2007 Population Census, the Nuer constitutes the majority of the population (40 per cent), followed by the Anywaa (27 per cent) and Majangir (5.8 per cent) (FDRE and Central Statistical Agency, 2007). It is worth noting that the Nuer constitute the second largest ethnic group in Ethiopia, next to the Dinka, among the Nilotes in South Sudan, while the Anuak also occupy the Pochalla district in South Sudan. The Nuer are predominantly pastoralist and the Anuak are sedentary agriculturalists, whereas the Majangir often combine hunting and gathering with shifting cultivation (Medhane, 2007).

The majority of the people in Gambella are dependent on land for farming, grazing and forest use (Figure 9.1). Since 2010, the regional state has invited potential investors to invest in high-value crops for export and import markets (MoARD, undated). The main rainy season of the region is from May to October with July and August as the most rain intensive period. The most important crops produced in the proposed investment areas are cotton, sesame, soybeans, groundnuts, rice, sugarcane, spices, vegetables and fruits. Local people have three farming systems: rain-fed, moisture and shifting cultivation. The Anywaa and Nuer farm twice a year using soil moisture farming. Out of the available land, only 0.5 per cent is suitable for moisture farming; the rest is, respectively, savannah 65 per cent, forestland 30 per cent and marshland 4.5 per cent (Monica, 2005).

Figure 9.1 Work on irrigation scheme in Gambella

Source: Ana Elisa Cascão.

Land

Land tenure in Ethiopia

During the Imperial regime (1929–1974), a mix of communal, church, state and private land tenure regimes existed. The *Derg* regime abolished private holdings in 1975 and land was nationalized and redistributed, giving peasants usufruct rights over land. The usufruct right did not include the right to sell, mortgage, exchange and lease. The hiring of wage labour was also prohibited during the *Derg* regime (PMAC, 1975, cited in Makki, 2012: 87). The land reform during the *Derg* regime 'established a relatively homogeneous smallholder agrarian order across the entire space of Ethiopia' (Makki, 2012: 85). The state monopolized the land tenure system and the government came to have plenipotentiary power to decide the type of land allocation and use (Scott, 1998, cited in Makki, 2012). In spite of the land reform, the agricultural growth rate steadily declined from 3.3 per cent in 1957–1961 to 2.9 per cent in 1963–1967, and further to 1.2 per cent in 1968–1973 (Clapham, 1988, cited in Makki, 2012: 87). Dessalegn (2009: 24) noted that 'the end product of the land reform was that it failed where it succeeded. It transformed rural Ethiopia into a society of self-laboring peasants.' The Gambella region has always remained at the peripheral end of the land reforms that was introduced by the *Derg* and studies assessing the impact of the land reform in Gambella is absent in the literature.

Since 1991, under the current Ethiopian Peoples' Revolutionary Democratic Front regime, land belongs to the people and the state. 'Policy has increasingly crystallized around a spatially differentiated promotion of smallholder commercialization and large-scale corporate agriculture' (Makki, 2012). Article 40 of the 1993 Federal Democratic Republic of Ethiopia (FDRE) Constitution states that 'the right to ownership of rural and urban land, as well as of all natural resources is exclusively vested in the state and the peoples of Ethiopia' (FDRE, 1993). Drawing inspiration from the developmentalist states in East Asia that 'had to base themselves on the explicit or tacit support of the peasantry' (Zenawi, 2006, cited in Makki, 2012: 90), the late Ethiopian Prime Minister Meles Zenawi explained the centrality of the existing land policy the following way:

> The advice we got from practically everyone was that we should privatize land … but we were not convinced that privatization of land was the way forward. Instead we opted to ensure tenure security for the peasantry and introduce a long-term land lease system for commercial farmers.
>
> (Zenawi, 2006, cited in Makki, 2012: 90)

Unlike the past, the current land tenure system has a direct impact upon the GRS. Up until the emergence of commercial farms in the Gambella region, the *de facto* communal landownership system was in practice. Water, cultivable land, grazing land, forest products and woodlands have been used communally among the indigenous people. The penetration of large-scale commercial farms in Gambella since 2010 has brought the traditional land-use system and the state's land tenure system to the centre of the debate. Some argue that the ongoing commercial farms replace the communal landownership system.

The Federal Rural Land Administration and Utilization Proclamations (FRLAUP) 456/2005, Article 5(3) states: 'Government being the owner of rural land can change communal rural land holdings to private holdings as may be necessary.' The practice of changing the ownership from communal land to private lease can be seen as an act of 'territorialization'[3] through legalization (Peluso and Lund, 2011). Many authors have noted that 'territorialization is another claiming and managerial practice long carried out by states and other entities' (Vandergeest and Peluso, 1995; see also Sassen, 2000; Peluso and Vandergeest, 2001; Brenner *et al.*, 2009, all cited in Peluso and Lund, 2011: 673), as it creates physical and economic enclosures. 'It is not only the special enclosure and acquisition, these processes usually lead to institutionalization of the land grabbing, through territorialization and legalization' (Peluso and Lund, 2011: 674).

Territorialization has been going on in the Gambella region for the past few years by the government, investors and the Nuer, who are encroaching in the Anywaa zone in search of grazing land and water (Dereje, 2006). The state territorialization has at times caused confrontations with the local communities' territorialization practices. By the same token, it can be said that in Gambella the investment farms' spatial enclosure and possession has helped the state to institutionalize control of land through territorialization and legalization. More often, 'the government has resorted to the notion that the land is *terra nullius*' and by so doing it does not go against the Constitution that 'explicitly protects the use rights of local inhabitants' (Makki, 2012: 96). Territorialization in Gambella has resulted in blocking the Nuer's transhuman movement of the customary route to the grazing land and water points. Moreover, in Itang special *woreda*, the only *woreda* that includes both Anywaa and Nuer, there are around 60 commercial agricultural farms that are blocking the Nuer's traditional trekking route for trade from Lare to the Gambella city (Wondwosen, 2016).

Land conception

The current land tenure debate 'homogenizes' the divergent land conceptions of the indigenous people. The Nuer and Anywaa have different conceptions of land use and control. Based on the state's understanding of 'free and used land', the pastoralist Nuer grazing land is perceived to be more 'vacant' than the Anywaa's farmland. The state has often been blamed for favouring farmers over pastoralists. Its land policy and development programmes[4] have been criticized for not acknowledging pastoralism as a legitimate way of life and livelihood. The non-existence of physical markers defining grazing land usually created problems for pastoralist communities whose pasture areas are often considered as 'no-man's land'. Moreover, the penetration of commercial farms has impacted on Anywaa–Nuer relations. The two groups have had acrimonious relations in the past. There have been many conflicts between the two groups, mainly in their attempts to use natural resources (land and water) and as a political space (Dereje, undated, 2007, 2009; Medhane, 2007).

The Nuer and Anywaa have contrasting land discourses. New large-scale land investments have further complicated the relations between the Nuer and the Anywaa.[5] The Nuer's land identity discourse is less territorialized than the Anywaa's. They adapt an 'open discourse' on the land. Their 'locus of identity is communitarian' (Dereje, 2006: 171). On the contrary, the Anywaa have an identity structured around land. They are highly territorial and deeply engaged with preserving their ancestral land. The Anywaa land narrative employs the moral sanction to territoriality by saying that 'one must be a bad person to leave his natal village'. Their conception of land has also 'spiritual dimension and supportive nature of the Earth' (Dereje, 2006: 171). For instance, 'resettlement in a new village requires an elaborate ritual of "befriending" the new earth' (Dereje, 2006: 171).[6] This notion of territoriality may pose challenges to land investment projects in the Anywaa districts. These divergent land conceptions between the two ethnic groups have further been described by Derje in the following way:

> The Anywaa notion of ownership does not entail effective occupation or possession. It is enough to put claims on once occupied territories, even without repossessing it. The Nuer recognize possession, which implies use right. Accordingly, if you do not use it, it does not belong to you. In this discourse ultimate ownership of natural resources is bestowed to Kuoth: the benevolent and patriarchal Nuer God, thereby discursively challenging the landed identity discourse of the Anywaa.
>
> (2006: 172)

Who owns the land in Gambella?

There are four major players in Gambella's land affairs: the indigenous people, the highlanders, the state and the investors. The Nuer and Anywaa are the two key ethnic groups that have been competing over the ownership of land and water resources in Gambella. As discussed above, the Anywaa and the Nuer radically differ in their conception of land and water. The Anywaa are territorialist and advocate an exclusionist land discourse, while the Nuer adhere to a more flexible and inclusive land discourse (Dereje, 2006). These two opposing land discourses together with the coming of large-scale agricultural land acquisitions has complicated the land and water nexus in the region. Human–land relations are in clear contrast in Gambella. Even if the Nuer comprise the majority of the population, the Anywaa own the majority of fertile farmlands, grazing lands and water resources in the region (Dereje, 2006) and most of the investment farms are located in Anywaa districts, where there are huge tracts of fertile land with rather low population density.

Nuer and Anywaa elites often employ different narratives and counter-narratives when claiming their ownership and entitlement to resources in the region. The Anywaa claim that they are the founders of the Gambella region, as the word Gambella is Anywaa. On the contrary, the Nuer use 'majoritarianism' as a counter-narrative to Anywaa's founding narrative of the Gambella region. The Anywaa contest the population census by invoking the 'citizenship' card, by labelling the Nuer as 'non-Ethiopians' who arrived in the Gambella region during the Sudan civil war (Dereje, 2006).

In a rather ingenious manipulation of state political discourse, the Nuer adapted the state's land tenure system in order to have access to land. On the other hand, the Anywaa have a more 'rigid' territoriality ideology towards land ownership. Dereje (2006: 256) noted

> the new discursive connectivity between the Nuer and the state's discourse on the land. Contemporary Nuer accrue ownership rights over the land to the state, while Anywaa territoriality is destined to disconnect with the state's own claim over the land. This is as much Nuer's exit strategy from Anywaa's closure as it is playing *realpolitik*.

However, 'such symbolic empowerment of the state is neither a commitment to the state nor an assertion of national identity … Nevertheless, it has had the effect of reproducing state ideologies at the local level' (Dereje, 2006: 172). Moreover, in an attempt to ridicule the 'Anywaa's exclusive land discourse which threatens to disown them as well, the Nuer often make an explicit statement that the land (the alluvial soil) after all comes from the highlands (Dereje, 2006: 172). This clearly strengthens the Anywaa's deep-rooted fear that the Nuer have always conspired with the state to displace them from their ancestral land (Dereje, 2006).

The pastoralist Nuer are considered to be expansionist while the agriculturalist Anywaa tend to have a static way of life. Evans-Prichard described Anywaa territoriality as follows: 'the Anywaa are strongly attached to the sites where their ancestors lived and often tenaciously occupied them in face of extermination' (cited in Dereje, 2010: 321). These contradicting views on property and territoriality seems to have been one of the major sources of conflict between the two groups. The Anywaa have occupied, as Dereje (2009) noted, the most fertile, riverine and strategic part of the Gambella region, occupying a great proportion of the landscape. On the contrary, the Nuer have occupied the smaller, degraded and less fertile areas. These age-old differences are further complicated by the advent of mega commercial farms in the region, which have strained the relations between these ethnic groups and complicated the question of landownership.

Large-scale agricultural land acquisition

The agricultural sector is the dominant sector in the Ethiopian economy and is estimated to contribute to about 47.7 per cent of the total gross domestic product (GDP), 85 per cent of employment and 75 per cent of the total export value (Mundi, 2013, cited in EEA, 2015: 39). There are around 12.6 million small farm[7] households in Ethiopia dependent on rain-fed agriculture (EEA, 2015: 55). The first and the second Growth and Transformation Plan (GTP)[8] aim to achieve a food-secure, middle-income country by 2025 and increase output of major crops (FDRE, 2010, 2015). The overarching policy and strategy for Ethiopia's agriculture is the Agricultural Development Led

Industrialization programme, which also envisages transforming the country into an industrialized nation (EEA, 2015). The government is trying to improve the capacity and efficiency of smallholder farmers in the highland and the centre, while it encourages large-scale agriculture at the peripheries and lowland regions such as Gambella.

Ethiopia has a huge cultivable land size, and is placed as the 27th largest country in the world[9] (Henok *et al.*, 2013: 225). The Ethiopian MoA (formerly Ministry of Agriculture and Rural Development, MoARD) stated that Ethiopia has 51.3 million hectares of agricultural land, of which only 11.7 million hectares, or about 20 per cent, are currently under cultivation. Deininger *et al.* (2011) estimated that about a total of 1.19 million hectares of land is leased out to large-scale farms in Ethiopia, of this 51 per cent of the land is owned by foreign investors. It has always been a serious problem to get accurate data about existing and potential agricultural land in Ethiopia, in general, and in Gambella, in particular. The land data discrepancy is also common when it comes to the amount of land the investors have acquired from the Gambella region. Opponents to the large concessions of land usually present exaggerated figures, while those advocating for the expansion of large-scale land investment often tend to downplay the negative impacts of these mega commercial projects.

The Gambella People Bureau of Economic Development estimated in 2000 that there were 129,014ha of cultivable land available and only 2.4 per cent was being cultivated (GPBED, 2000, cited in Dereje, 2006). After 2010, the area allotted for commercial farming has skyrocketed. In a preliminary assessment, the GRS identified about 1,238,000ha as a potential area for large-scale commercial farming in 12 districts in the Gambella region (MoARD, undated). In 2009, the Government of Ethiopia identified 3.7 million hectares of land suitable for agricultural investments, of which 444,150ha were identified in the Gambella region and placed under the federal government land bank[10] (REGLAP, 2011). A report from the MoARD states that Gambella has a potential of 200,000ha only for the production of corn using rain-fed agriculture and irrigation (MoARD, 2009).

One of my informants from the MoA had a different take on this. According to this informant, what has not been properly accounted for is that out of 263 investors in Gambella, only a dozen of them are international investors, as can be seen in Table 9.1. The potential contribution of local investors for the country's food self-sufficiency and input for the industry (particularly cotton) has always been downplayed by those who oppose land acquisition. A domestic investor said that 'in the past the rich used to be blamed for crowding the service sector, specifically the catering and hotel services, and now we are blamed for venturing into farming. There are fierce critics on us even before we start farming.' Then he asks:

> Which sector do you think is easy for me: building a hotel in Addis Ababa or farming in Gambella? I am telling you, security wise, I prefer to have a

> hotel in Addis, but sometimes you respond to the country's call and make your input in the manufacturing industry, which is now heavily dependent on Chinese products.[11]

Currently, there are about a dozen international investors in Gambella (see Table 9.1). Indian investors have dominated the overall investment landscape in the region, followed by Saudi, Turkey and China investors. Gambella is the only region where the largest land size was given out to investors. Karaturi, an Indian investor, had 100,000ha of land before the land lease agreement contract was terminated and the investment licence revoked in 2016 due to the inability to develop the land.

Table 9.1 International investors in Gambella

	Nationality	*Investment type*	*Land transfer area in hectares*	*Capital registered / Mill Birr*	*Land rent*	*Receipt number*	*Agreement signed date*
1	India	Cotton and soya bean	25,000	436	3,950,000	331025	02/09/2011
2	Turkey	Cotton and soya bean	6,000	1,000	948,000	531378	12/01/2012
3	India	Cotton farming and related activities	5,000	17	790,000		25/06/2012
4	India	Cotton farming and related activities	5,000	17	555,000		25/01/2012
5	India	Soya bean	25,000	1,451	2,775,000		27/07/2010
6	India	Edible oil crops	27,000	918	2,997,000	136702	03/09/2010
7	India	Rice	10,000	160	1,580,000	141636	24/01/2011
8	India	Tea	3,012	631	334,332	259361	13/08/2010
9	India	Palm cereals, rice and sugar cane	100,000	2,110	2,000,000		26/02/2011
10	Saudi	Rice	10,000	37,640	300,000		22/02/2011
11	China	Sugar cane	25,000	2,973	3,950,000	192498	05/11/2010

Source: collected from the Agricultural Investment Support Directorate (AISD), MoA – now renamed and expanded as an independent body: the Agricultural Investment Land Administration Agency (AILAA).

Centering the periphery

The Ethiopian state has always been represented by the highlanders who comprise the 'significant others' in contemporary Anywaa–Nuer relations (Dereje, 2006: 242). Currently, for the local people, the federal government, the investors and the highlanders are one and the same. Economically speaking, the highlanders benefited from their control of the exchange and commercial sector, ranging from petty to wholesale trade. For instance, out of 'the 691 officially licensed traders only a handful are Anywaa and Nuer' (Dereje, 2009: 10). The data from the investment office also show that all the diaspora and domestic land investors are highlanders. The indigenous people consider large-scale agricultural investments as a continuation of the highlanders' political and economic dominance and the state's thirst for more resources. The advent of commercial farms in the region has led to new political dynamics in the highland-centre and lowland-periphery interaction. There is now more central government involvement through both institutional and legal control and through the influx of the highlanders for jobs and commercial ventures.

The centre and periphery relationship have always been a vexed one. As Makki (2012: 85) puts it: 'The general developmental dynamic of the Ethiopian social formation was rooted in this divergent core-periphery structure.' He further noted that 'scholarly and popular images of Ethiopia have been profoundly shaped by a pervasive geographical contrast between highlands and lowlands' (2012: 84). This divide has put the 'civilization discourse juxtaposing the *mahel-ager* (the center) to the *dar-ager* (the periphery)' (2012: 85). Moreover, '[in] the *longue dureé* the geographical divide consequently served as a socio-cultural and administrative divide' (see Donham and James, 1986; Markakis, 1998, both cited in Makki, 2012: 85). This core–periphery relationship can be summarized as follows:

> Frontiers are not sites where 'development' and 'progress' meet 'wilderness' or 'traditional lands and peoples'. They are sites where authorities, sovereignties, and hegemonies of the recent past have been or are currently being challenged by new enclosures, territorializations, and property regimes. What is new is not only land grabbing or ownership but also new crops with new labor processes and objectives for the growers, new actors and subjects, and new legal and practical instruments for possessing, expropriating, or challenging previous land controls.
>
> (Peluso and Lund, 2011: 668)

The centre–periphery interaction has always been hierarchical, where the periphery is located at the lower and weaker end of this relation. This correlation is now being played out in the large-scale agricultural investment area. To borrow Coronil's (2001: 67) words, the strong–weak center periphery relations 'reproduce asymmetric power relations' between them. Some argue that the land deals serve and advance the interest of the central government and give it 'an opportunity for extending its reach, exerting power over marginal areas and peoples, as well as extracting rent from such "unruly" places' (cf. Scott 2009, cited in White *et al.*, 2012: 627). Corporate land grabbing is thus linked to a new expansion of state authority and control' (Borras *et al.*, undated, cited in White *et al.*, 2012: 627).

Intra- and inter-border competition over land and water

Deep-rooted intra- and inter-border conflicts between the Nuer and Anuak over land and water resources have been common in Gambella over the past years. There are different drivers behind the conflicts in the region. As discussed above, one of the major reasons has been the struggle to control scarce natural resources such as water and land, and conflict arising from the resistance of local peoples to the political and demographic domination of those who had come from the highlands. There has been and still is a general feeling of apprehension among the Anuak that they are being outnumbered and dominated by the pastoralist Nuer who enter their territory from Sudan in search of grazing land and water. It is closely linked to the problem of weak security in South Sudan that impacts upon and spills over into the Gambella region. The movement of the Nuer into Gambella has resulted in conflicts with the Anuak (in Gambella), Dinka (in South Sudan), and even among the Nuer themselves who inhabit areas in both Gambella and South Sudan.[12] To use Medhane's (2007: 7) words, 'the expansionist nature of the Nuer for grazing land and drinking water and the Anywaa's' strong emotional ties with their land' are the primary sources of conflict. It is argued that the situation is increasingly conflictual given the upsurge in land leasing.

Gambella is in many ways linked to what is happening in South Sudan. Any discussion of the Gambella region cannot be comprehensive without factoring the relationships it has with South Sudan. What goes on in South Sudan has a direct impact in Gambella and vice versa. The typical example in this regard are the effects that the Sudan and South Sudan civil wars have had on Gambella. Similar to Gambella, there has been massive land leasing in South Sudan as well. There are '28 large-scale investments in South Sudan in agriculture, biofuel or forestry covering 2.64 million ha, signed between 2007 and 2010 by both domestic and foreign investors' (Mosley, 2012). This process may create more potential and real conflicts among the major ethnic groups between the Nuer and Anuak. As Medhane (2007: 7) perceptively said, 'ethnic boundary is … marked by difference in livelihood strategies'. Conflict in the region is thus a function of differences 'between two socio-cultural groups and was related largely to sedentary (Anuak) and pastoral (Nuer) resources' (Medhane, 2007: 7). A quick retrospective look reveals that 'the Nuer and Anuak were the pawns in the "cold war" between Sudan and Ethiopia' (Medhane, 2007: 8).[13] The civil war in South Sudan, land leasing in the two countries and the political economy of the region have impacted upon the rivalries between the Nuer and the Anuak in Gambella and South Sudan.

Water

Ethiopia is one of the richest countries in Africa in terms of water resources. Ethiopia has 12 major river basins, with the total mean annual flow of 122bcm (billion cubic metres) (MoWR, 1999, cited in Awulachew *et al.*, 2007). The irrigation potential[14] is estimated at 4.3 million hectares, of which only 1 million is being utilized (MoARD, 2009). It is often referred as the Water Tower of the Horn of Africa. However, with the looming of climate variability and climate change, the accessibility and reliability of water remains challenging.

Baro-Akobo Basin

The Baro-Akobo-Sobat Basin is located in the Gambella region, on the Ethiopia–South Sudan border. It consists of four major rivers, namely Baro, Alwero, Gilo and Akobo. It has a total catchment area of nearly 76,000km^2 with an annual runoff of 23.24bcm (that is about 21 per cent of the country's total annual surface flow; see Table 9.2). The basin has a lowest elevation of 390 metres and highest elevation of 3,244 metres. It comprises 28.75 per cent of the catchment area and 21.2 per cent runoff of the Nile Basin in Ethiopia. The Baro-Akobo is called Sobat after crossing the South Sudan border, before it confluences with the White Nile River. The Baro-Akobo river system has a huge irrigation and hydropower potential (MoWR, 1997; NBI, 2005).

The Baro-Akobo-Sobat Multipurpose project was one of the eight integrated water resource development projects designed by the Nile Basin Initiative (NBI) under the Eastern Nile Subsidiary Action Programme (ENSAP) to be implemented in the Gambella region. The project aimed to

Table 9.2 The Baro–Akobo River Basin

Number	*The Baro–Akobo River Basin and its sub-basins*	*Catchment area (km^2)*	*Mean annual runoff (bcm)*
1	Baro	30,004	12.78
2	Akobo Upper	6,036	1.77
3	Akobo Lower	7,209	2.12
4	Gilo	12,815	3.22
5	Alwero	8,019	1.38
6	Serkole	7,702	1.32
7	Tirmatid	2,690	0.42
8	Pibor	1,435	0.22
	Total	**75,910**	**23.24**

Source: Awulachew *et al.* (2007).

harness hydropower generation of up to 793 megawatts (MW), irrigate about 20,000ha, and regulate the river for conservation, sediment and flood control through the construction of five dams on the Baro River and two dams (one on the Sobat and one at the Akobo-Pibor confluence) (ENTRO, 2006). This project, however, was never realized due to the Sudan civil war and a lack of sufficient funding.

Water conception

The Ethiopian government and the Ethiopian people usually see Ethiopia as a water tower, not as a water grabber. Given the fact that almost all of Ethiopia's rivers are transboundary rivers, Ethiopia has to utilize its water resources to aid its development projects and to improve the economic life of its population. Even if Ethiopia is considered as a water tower of the Horn of Africa, it has never utilized its water resources significantly. Thus, there is a strong commitment to develop its water potential. The country employs a narrative of harnessing its water resources for poverty alleviation and food security, as a development imperative. As an informant from the Ministry of Water, Irrigation and Electricity noted, boundary and transboundary water resource utilization is 'a noble and patriotic gesture to eradicate food insecurity and famine'[15] that has tarnished the country's image for decades.

The same informant noted that 'whoever uses the water resources in Ethiopia can hardly be called a "water grabber" or as engaged in "privatized water" acquisition'. Such an act, my informant stated, should be considered as an attempt to tame nature and resources at ones disposal. 'Damming the rivers is not grabbing water, but making the best out of ones resources.' 'Developmental hero' is an appellation often used to describe the late Prime Minister, Melese Zenawi, who in his advocacy for a 'democratic developmental state' called for fighting chronic poverty, which is said to be 'Ethiopia's imminent enemy'. The government annually awards farmers who have proved to demonstrate outstanding effort all over the country on live television and radio transmission programmes with a decorated Ethiopian flag medal, acknowledging them as 'developmental heroes' who increased land and water productivity and for contributing for the growth of agricultural production in the country. By the same token, utilizing the Baro-Akobo River can be seen as yet another testimony of the country's firm determination to extricate itself from the quagmire of poverty and achieving food self-sufficiency and food security.

In Ethiopia water is abundant, but land is increasingly becoming scarce. Apart from actual precipitation and climate variabilities, water availability is unevenly distributed and accessible. Land is considered a very crucial property while water is 'God Given'. Land is priceless while water is free. Throughout Ethiopian history, land has been regarded as the key resource and continues to have an important place in the political history of the country. The 1974 revolution was fuelled by the land question, which resulted in the overthrow of the emperor. Water, on the other hand, has not been a central element in the country's politics until recently, and it is only the use of the Nile waters that has attracted a lot of political attention in the recent history of the country. The main conception of water among Ethiopians seems to be to use as much water as one needs, and what is holding people back from using more water is the capacity to utilize our rivers.

Land Lease Agreement and water use

Water use by land investors (both international and national) is generally regulated in the contracts by reference to Ethiopia's Land Lease Agreement (Figure 9.2). Article 3, Rights of the Lease, sub-section 3.2, stipulates that the lessee has the right to:

> Build infrastructure such as dams, water boreholes, power house, irrigation systems, roads … at the discretion of the Leasses upon consultation and submission of permit request with concerned offices subject to the type and size of the investment project whenever it deems so appropriate.

As can be seen from this article, water use is grouped under the broader 'infrastructure development'. It is also interesting to note that the clause 'upon consultation and submission of *permit* request with concerned offices' shows that there is a need to get permission to develop the water resources infrastructure. But the article does not place any price for water or irrigation. Rather, it provides a right for the investor to develop the water as deemed necessary if they obtain the permit from the relevant offices.

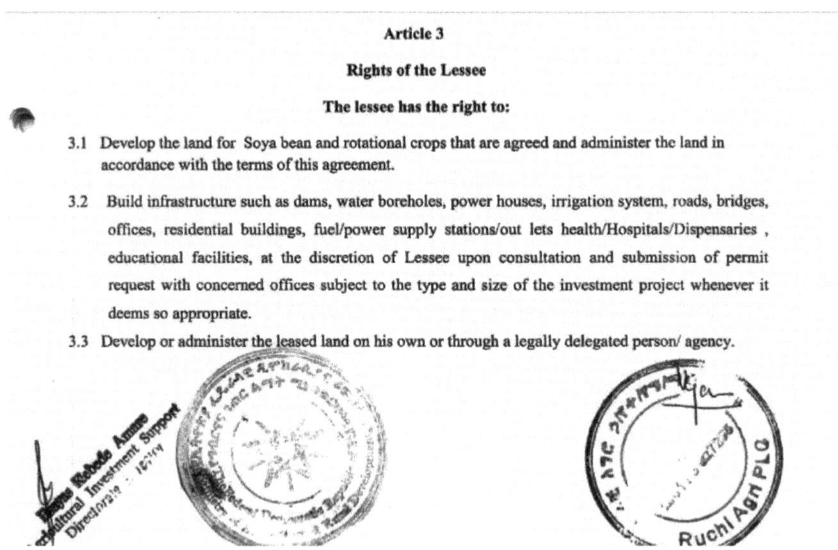

Article 3

Rights of the Lessee

The lessee has the right to:

3.1 Develop the land for Soya bean and rotational crops that are agreed and administer the land in accordance with the terms of this agreement.

3.2 Build infrastructure such as dams, water boreholes, power houses, irrigation system, roads, bridges, offices, residential buildings, fuel/power supply stations/out lets health/Hospitals/Dispensaries , educational facilities, at the discretion of Lessee upon consultation and submission of permit request with concerned offices subject to the type and size of the investment project whenever it deems so appropriate.

3.3 Develop or administer the leased land on his own or through a legally delegated person/ agency.

Figure 9.2 Land Lease Agreement

Source: land rent contractual agreement between MoA and BHO Bio Product PLC, Land Leasing Agreement (2010).

This unlimited and unregulated water use permit may create pressure on water resources that has implications on downstream users. Unregulated water use does not run in tandem with the principles of integrated water resources management and development. Considering the looming challenge of climate change, these uncoordinated usages of water resources will affect agricultural productivity and water security in the region. The question remains: does this article promote water privatization/enclosure or water resources development? The vagueness of the article neither discourages 'water privatization by investors' nor ensures the 'water use rights of the local community'; rather, it falls under the wider 'infrastructure development' narrative, which is much broader than the water use or grab *per se*. Other studies (Smaller and Mann, 2009; Cotula, 2011; Woodhouse and Ganho, 2011, cited in Jägerskog *et al.*, 2012: 6) have found that the land leasing agreement both glosses over the 'water use and rights' of local communities and gives 'subtle and implicit' 'water use rights' to investors.

Water use/grab

Land investment/grab is a water investment/grab (Oakland Institute, 2011; GRAIN, 2012; Jägerskog *et al.*, 2012: 6). When land is leased out in Gambella, water is also implicitly given out free of charge. This process transforms water from 'resources openly available' to 'private or an investor's goods'. As a result, access to water resources may become difficult for many communities in the Gambella region. The ongoing, large-scale agricultural projects have resulted in the 'enclosure of the water resources', preventing water access for the poor and marginalized groups. Land investments not only affect water quantity, but also water quality, which has been overlooked in the debate of water grab. The large and numerous agricultural farms not only use water resources, but also pollute the water. Consequently, the regulation of water use permit/rights should also incorporate the 'water quality' dimension, as the agrochemicals on the farm pollute the water, hence, reducing the productive nature of water.

Water and land (in)security

Both water and land are considered to be in abundance in the Gambella region. Nonetheless, the introduction of large-scale land investment in the area may significantly reduce the amount of available water. An informant from Addis Ababa University argued that the current wave of land acquisitions has created land and water insecurity in Gambella, and that the availability of both grey and green waters is becoming a problem in the region. In addition, the same informant indicated that 'property insecurity' has been the main issue in the region. The Nuer fear losing grazing land and water points and the Anywaa are concerned about losing their farmland with the introduction of large-scale land investments in the area.[16] Generally, changes in property rights often tend to involve winners and losers (Alston *et al.*, 1996) and the impact of commercial farms on small farms' productivity still awaits detailed research.[17] Food is at the centre of water and land dynamics in Ethiopia. The same informant did not hesitate to state that the ongoing land acquisition in Gambella will compromise the productivity of smallholders and may result in household food insecurity. An expert in the MoA completely disregarded this claim and argued that commercial farming will benefit smallholders in the area through knowledge transfer and the provision of fertilizer at cheap prices, etc. In responding to critics of commercial farms in Gambella, the same expert from the MoA noted that 'they have empty pages to fill, but we have empty stomachs to feed … How can we feed our people without using large-scale agriculture, since we know that small scale farming "remains small" to feed our people?'[18]

Flood and land

Gambella is prone to two contrasting natural calamities, drought and floods, causing human and material losses. As a result, the central government and the Gambella regional administration has used this as an argument to design a 'villagization' programme to engage the sparsely populated community settled in the riverside in shifting cultivation, and to move those exposed to natural disaster to safer and better settlement sites from 2010/2011 to 2013/2014 (Gambella Regional State, 2010). The flood from the river is, however, vital for moisture farming and dry-season pasture. An informant, who works in the Itang *woreda* Agricultural Bureau, noted that with the arrival of large-scale commercial farms, there were attempts to regulate the flood by, for instance, BHO, an Indian investor. The local community and households in the *woreda* have mixed feelings about flood regulation. Households who previously were affected adversely by the flood were generally pleased that the flooding now had ceased, whereas those who were using the flood for receding cultivation and dry-season pasture generally expressed strong concerns about the villagization project. The Nuer harvest maize and sorghum twice a year using the flood waters. They also practise transhumance between wet season villages and dry season camps by following the ebb and rise of the Baro River flow. According to an informant from the Agricultural Bureau, generally in the Gambella region, the Nuer, who mainly use moisture agriculture and dry-season grazing land, are more susceptible to the reduction of the floodwaters than the Anywaa who practise rain-fed agriculture. An elderly person in Itang *woreda* wisely said: 'Flood is like a snake. You get medicine out of a snake, just like we make our livelihood by using the flood. All you have to know is how to handle it.'[19] While some of the indigenous people have a bittersweet experience of flood, the investors on the contrary were unprepared for the devastations caused by flooding. For instance, Sai Ramakrishna Karuturi, CEO of Karuturi Global who lost US$15 million, said that '[t]his kind of flooding we haven't seen before. This is a crazy amount of water.'[20]

Transboundary water management and land investment

Generally, the issue of large-scale investment in Gambella has not been a main concern among downstream countries. Given the fact that the Baro-Akobo-Sobat Basin only contributes to a small portion of the main Nile waters, there is little or no concern from downstream countries, at least for the time being. Another reason might be that Gambella is a rain-rich area where much of the cultivation is rain-fed. In Gambella, both small-scale and large-scale domestic and international investors are mainly relying on rain-fed agriculture. 'Globally, rain-fed agriculture is the most common practice, especially in developing countries' (Jägerskog *et al.*, 2012: 10). In sub-Saharan Africa, 96 per cent of the cultivated land is not irrigated but rain-fed (FAO AQUASTAT, 2012, cited in Jägerskog *et al.*, 2012: 10). Another factor are the conflicts in South Sudan, which have hindered the country from developing water resources on a large scale.

The other important factor is that it has always been the Blue Nile, not the Tekeze or Baro-Akobo River, that has been at the centre of Nile water discussion and dispute. The Baro-Akobo River has always been out of the hydropolitical radar. In this regard, it is not surprising that the issue of Baro-Akobo in Gambella has never surfaced in the Nile water discussions or hit the headlines of the newspapers. Most of the contemporary reporting in the Gambella focuses on 'land grab' and not on 'water grab' in the Nile Basin and Horn of Africa region. As it stands now, the Baro-Akobo River is not considered as an important site of transboundary politics in the region. Nonetheless, this may not remain so if water withdrawal becomes significant in the future.

Land investment in Gambella has brought an interesting element in the use of water resources in the area. At the national and regional level, Ethiopia employs a 'reasonable and equitable' principle of water sharing, while at the local level, these principles are 'overlooked' by a pragmatic land investment approach. Water allocation, amount and use are completely left at the discretion of the 'private' investor, while at the country and regional level, water allocation is a national issue. As such, at the local level water is an 'issue of private investor' while at the country and the NBI level 'it is an issue of national decision'. The two approaches are paradoxical. New land deals or investment are beyond the 'mandate of the NBI'. To date, the NBI has not issued any guideline or strategy in the use of Nile waters in relation to large-scale commercial farms. This reinforces the often-mentioned unilateral development approaches undertaken in the Nile Basin. Land leasing in Gambella can be seen as one instance of such 'unilateral' development project. In fact, almost all Nile riparian countries encourage individual domestic and international investors to invest in land in the basin.

Conclusion

It is possible to observe several incongruent stories about current land and water acquisition in the Gambella region, ranging from perceptions that 'the government is going to abolish the local community's livelihood' and 'highlanderize Gambella', to those who spoke about the government's work as a 'genuine step to fight poverty and food insecurity'. These contradictory views have, to an extent, emanated from a lack of proper government consultation with local communities and the 'opaque' nature of the land lease agreements. Considering the strained historical relationships between the Gambella region and the government, it is not surprising to hear such contradictory stories about the ongoing land and water acquisition in Gambella.

The entering of large-scale agricultural land acquisitions in the Gambella region has also contributed to further straining the relationship between the Anywaa and Nuer vis-à-vis the government and the land investors, which may lead to future disputes among the ethnic groups, investors and the government.

Even though water and land are considered to be in abundance in Gambella, the introduction of large-scale land investment will put a big question mark on the amount of available land and water resources. 'Property insecurity' has been the main issue of concern in the region. Whenever the land is being leased out in Gambella, water is also implicitly given out free of charge and unlimitedly. This process may transform water from 'resources openly available' to 'a private or an investor's goods'.

Ethiopia has utilized a meagre amount of its immense water potential. Today, there is a commitment and aspiration to harness both boundary and transboundary waters. The state narrative, which is widely subscribed by Ethiopians, is to see 'Ethiopia as a water tower, and potentially a water user, not a water grabber'. An analysis of the Land Lease Agreement shows that investors or lessees need to have a 'water permit' to use the waters for agriculture. The agreement has a vague position as it neither discourages 'water privatization by investors' nor ensures the 'water use rights of the local community'. The agreement, furthermore, does not put any limitation as to 'how much water, when, and where' to use the water and rivers in Gambella.

Notes

1 In this chapter, I use 'large-scale agricultural land acquisition' to avoid the assumption that every land acquisition is a 'land grab' and to avoid using the notion of 'land investment' that tends to downplay the political implications of land acquisition. Cotula (2010: 1) noted that '[n]ot every acquisition is a "land grab" or a development opportunity'.

2 Namely Tigray, Afar, Amhara, Oromia, Somali, Benishangul-Gumuz, Southern Nations Nationalities and People Region (SNNPR), Gambella and Harari, and two administrative cities: Addis Ababa City administration and Dire Dawa City Council.

3 Several contributors show that under neoliberal economic policies, the mechanisms by which state territorialization takes place involve a variety of legal instruments and institutional alliances and agreements between state, non-state and parastatal institutions

(Corson, 2011; Lund, 2011; Osborne, 2011; Woods, 2011; Chapin, 2004, all cited in Peluso and Lund, 2011).

4 The Poverty Reduction Strategic Papers (PRSP I & II) (1994/1995–2004/2005), the Plan for Accelerated and Sustained Development to End Poverty (PASDEP) (2005/2006–2009/2010) and the first GTP (2010/2011–2014/2015) and the second GTP (2015–20120) have either completely glossed over or partially touched the issue of pastoralism.

5 Dereje (2007) notes:

> Nuer expansion at the expense of Anywaa territories has taken three forms: (1) the nineteenth-century conquest; (2) land encroachments by resource poor Nuer sections (particularly in Jor and Akobo areas), and (3) the recent influx to the towns to have access to social services most of which are located in the traditionally Anywaa territories including the regional capital, Gambella town.

6 The Anywaa's fiercer refusal than Nuer's to the government's large-scale agricultural acquisitions may be explained from the Anywaa's notion of land discourse.

7 Based on the Ministry of Water Resources (MoWR) classification, irrigation projects in Ethiopia are identified as large-scale irrigation (LSI) if the size of command area is greater than 3,000ha, medium scale if it falls in the range of 200ha to 3,000ha and small scale if it covers less than 200ha (cited in Awulachew *et al.*, 2007).

8 The first GTP was from 2010/2011 to 2014/2015 and the second GTP is in the pipeline.

9 In terms of availability of uncultivated land, sub-Saharan Africa stands first with a potential of more than 200 million hectares, followed by Latin America and the Caribbean with available land of about 123 million hectares (Deininger *et al.*, 2011). A recent finding by Deininger *et al.* (2011) reported that in a period of less than one year, foreign investors expressed their interest to lease in 29 million hectares of land from sub-Saharan Africa. The World Bank (2010, xiv) notes that 'investors expressed interest in 42 million ha of land globally in less than a year [between 1 October 2008 and 31 August 2009]' (cited in De Schutter, 2011).

10 The land bank identified 409,678ha in the Awash River Basin, 180,625ha in South Omo and 691,984ha in Benishangul-Gumuz (REGLAP, 2011).

11 Personal communication.

12 For example, there was a conflict between the two different sections of the Nuer, (Jikany and Lou), starting from 1991; for further discussion, see 'Peace and Conflict among the Nuer', www.pastoralists.org/pages/posts/peace-and-conflict-among-the-nuer9.php (accessed October 2015).

13 The following examples would help to show the relationships between Gambella and South Sudan : 'Influx of Sudanese Nuer refugees further tilted the balance of power in favour of the Nuer in Gambella' and another incident was 'The Anuak became increasingly isolated as the Nuer, SPLA and *Derg* formed a loose coalition against the Anuak' (Medhane, 2007: 8, 9).

14 FAO (2005) estimated 3.6 million hectares of irrigation potential.

15 At the time of writing this chapter, Ethiopia is experiencing one of the worst El Niño-driven drought in decades. According to the Humanitarian Requirements Document (HRD) for 2016, issued on 11 December jointly by the government and donors, 10.2 million people will be in need of food emergency assistance this year (Ethiopia: Humanitarian Overview, 2016).

16 This is, in fact, not limited to the Gambella region; in its enormous infrastructural development the government has been displacing people for road construction and expansion, from Addis Ababa to the periphery. This has been going on by partial citing of the constitution, which clearly stated that everyone can be displaced for the development sake, with proper compensation. Privatize everything? In its drive to 'privatize everything' (Watts, 1994).

17 Previous studies on the link between agricultural productivity and landholding security have opposing views. For instance, Deininger (2003) notes 'property rights affect economic growth in a number of ways'. On the contrary, studies made by Migot-Adholla *et al.* (1991: 172), in sub-Saharan Africa (Ghana, Kenya and Rwanda) state 'regression analysis indicates no relationship between cross-sectional variations in land rights and productivity' (cited in Shimelles *et al.*, 2009).
18 Personal communication.
19 Ibid.
20 See http://farmlandgrab.org/post/view/19411 (accessed December 2015).
21 Ethiopian authors name are written by their first name, followed by their father's name as Ethiopians do not follow or use their last name.

References[21]

Alston, L. J., Eggertson, T. and North, D. C. (1996). Toward an understanding of the property right. In Alston, L. J., Eggertson, T. and North, D. C. (eds). *Empirical Studies in Institutional Change*: 31-33. Cambridge: Cambridge University Press.

Awulachew, S. B., Yilma, A. D., Loulseged, M., Loiskandl, W., Ayana, M. and Alamirew, T. (2007). *Water Resources and Irrigation Development in Ethiopia*. International Water Management Institute (Working Paper 123). Colombo: IWMI.

Coronil, F. (2001). Toward critic of globalceterism: Speculation of capitalism's nature. In Comarotte, J. and Comoroffuke, J. L. (eds). *Millennial Capitalism and the Culture of Neoliberalism*: 63–87. Durham and London: Duke University Press.

Cotula, L. (2010). *Investment Contracts andSustainable Development: How to Make Contracts for Fairer and More Sustainable Natural Resource Investments*. Natural Resource Issues No. 20. London: IIED.

Cotula, L. (2011). *Land Deals in Africa: What is in the Contracts?* London: IIED.

Cotula, L., Vermeulen, S., Leonard, R. and Keeley, J. (2009). *Land Grab or Development Opportunity? Agricultural Investment and International Land Deals in Africa*. London/Rome: IIED/FAO/IFAD.

Deininger, K. (2003). *Land Policies for Growth and Poverty Reduction*. A World Bank policy research report. Washington, DC: World Bank Group.

Deininger, K., Byerlee, D., Lindsay, J., Norton, A., Selod, H. and Stickler, M. (2011). *Rising Global Interest in Farmland: Can it Yield Sustainable and Equitable Benefits?* Washington, DC: World Bank.

Dereje Feiyssa (2006). *Differing Identity Discourses, Differing Ethnicities: Comparing the Anywaa and the Nuer*. Berlin: Max Planck Institute for Social Anthropology.

Dereje Feiyssa (2007). *The Interface between National and Regional Level Decisions and Local Conflicts in the Gambella Region*. Research report submitted to Pact Ethiopia's Restoration of Community Stability in Gambella Project and Constructive Dialogue Initiative Project.

Dereje Feiyssa (2009). *The Current Conflict Situation in the Gambella Region and Suggestions for Sustainable Peace-building, With a Special Reference to the Baro River*. Research report submitted to Pact, Ethiopia.

Dereje Feiyssa (2010). The cultural construction of state borders: The view from Gambella. *Journal of Eastern African Studies*, 4, 2: 314–330.

Dereje Feiyssa (undated). *Taking the Bull by its Horns: Expanding the Natural Resource Base and Tackling the Root Causes of Conflict*. The Netherlands: Pax Chrisit.

De Schutter, O. (2011). How not to think of land-grabbing: three critiques of large-scale investments in farmland. *Journal of Peasant Studies*, 38, 2: 249–279.

Dessalegn, Rahmato (2009). *The Peasant and the State: Studies in Agrarian Change in Ethiopia 1950–2000s*. Addis Ababa: AAU Press.

Dessalegn, Rahmato (2011). *Land to Investor: Large-Scale Land Transfers in Ethiopia*. Forum for Social Studies.

Eastern Nile Technical Regional Office (ENTRO). (2006). *Eastern Nile Subsidiary Action Program Status Report, April 2006* (unpublished).

Ethiopia: Humanitarian Overview (2016). Joint Government and Humanitarian Partners Document. Synopsis.

Ethiopian Economic Association (EEA). (2015). *Report on the Ethiopian Economy. Prospects and challenges of Structural Transformation in Ethiopia*. Addis Ababa, Ethiopia.

FAO (2005). *FAO Water Report 29*. Brussels.

FDRE (1993). *The Federal Democratic Republic of Ethiopia Constitution 1993*. Addis Ababa.

FDRE (2010). *The First Growth and Transformation Plan (GTP) 2010*. Ministry of Finance and Economic Deelopment (MoFED), Addis Ababa.

FDRE (2015). *The Second Growth and Transformation Plan (GTP) 2015*. Ministry of Finance and Economic Development (MoFED), Addis Ababa.

FDRE and Central Statistical Agency (CSA) (2007). *Population and Housing Census Report Ethiopia*. Addis Ababa.

Franco, J., Mehta, L. and Veldwisch, G. J. (2013). The global politics of water grabbing. *Third World Quarterly*, 34, 9: 153.

Franco, J., Feodoroff, T., Kay, S., Kishimoto, S. and Pracucci, G. (2014). *The Global Water Grab: A Primer*, www.tni.org/en/publication/the-global-water-grab-a-primer (accessed December 2015).

Gambella Regional State (2010). *Villagization Programme Implementation Manual 2010*. Addis Ababa, Ethiopia.

GRAIN (2008). Seized! The 2008 land grab for food and financial security. *GRAIN Briefing*, October.

GRAIN (2012). *Squeezing Africa Dry: Behind Every Land Grab is a Water Grab*, 11 June.

The Guardian (2011). Africa's great 'water grab', www.guardian.co.uk/global-development/poverty-matters/2011/nov/24/africa-water-grab-land-rights?INTCMP=SRCH (accessed January 2016).

The Guardian (2015). Ethiopians talk of violent intimidation as their land is earmarked for foreign investors, www.theguardian.com/world/2015/apr/14/ethiopia-villagisation-violence-land-grab (accessed November 2015).

Henok, A., Bienen, D. and Ciuriak, D. (2013). Ethiopia's investment prospects: A sectoral overview. *African Review of Economics and Finance*, 4, 2: 204–246. African Centre for Economics and Finance.

Human Rights Watch (2012). *Waiting Here for Death: Forced Displacement and Villigization in Ethiopia's Gambella Region*. New York: Human Rights Watch.

Jägerskog, A., Cascão, A., Hårsmar, M. and Kim, K. (2012). *Land Acquisitions: How Will they Impact Transboundary Waters?* Report No. 30. Stockholm: SIWI.

Makki, Fouad (2012) Power and property: Commercialization, enclosures, and the transformation of agrarian relations in Ethiopia, *Journal of Peasant Studies*, 39, 1: 81–104.

Medhane, Tadesse (2007). *Gambella: The Impact of Local Conflict on Regional Security*. South Africa: Institute of Security Studies, ISS and Centre for Policy Research and Dialogue, CPRD.

Mehta, L., Veldwisch, G. J. and Franco, J. (2012). Introduction to the Special Issue: Water grabbing? Focus on the (re) appropriation of finite water resources. *Water Alternatives* 5, 2: 193–207.

Ministry of Agriculture (MoA) and BHO Bio Product PLC Land Leasing Agreement (2010). Addis Ababa.
Ministry of Agriculture and Rural Development (MoARD) (2009). *The 2009 Agricultural Potential Report of Ethiopia*. Addis Ababa.
Ministry of Agriculture and Rural Development (MoARD) (undated). *Invest in Ethiopia, Gambella Regional States*. Addis Ababa.
Ministry of Water Resources (MoWR) (1997). *Baro-Akobo River Basin Integrated Development Master Plan Project, Executive Summary*. Addis Ababa: FDRE/MoWR.
Mittal, Janhavi (2013). *Report from the Indian-Ethiopian Civil Society Summit on Land Investments*. New Delhi: Oakland Institute, 5–7 February.
Monica, M. S. (2005). *Transforming Conflict in the Multi-Ethnic State of Gambella in Ethiopia: Mediation in a Divided Society*. Bramois, Switzerland: Institut Universitaire Kurt Böschs.
Mosely, J. (2012). *Peace, Bread and Land: Agricultural Investments in Ethiopia and the Sudan*. London: Chatham House.
Nalepa, Rachel (2013). Land for agricultural development in the era of 'land grabbing': A spatial exploration of the 'marginal lands' narrative in contemporary Ethiopia. *Land Deal Politics Initiative*, Working Paper 40.
Nile Basin Initiative (2005). *Nile Transboundary Micro-grants Program, National Action Plan for Ethiopia (NBI/NTEAP)*. Prepared by the Local Micro Grant Coordination Office. Addis Ababa.
Oakland Institute (2011). *Understanding Land Investment Deals in Africa Country Report: Ethiopia*. New Delhi.
Oakland Institute (2013). *Unheard Voices: The Human Rights Impact of Land Investment on Indigenous Communities in Gambella*. New Delhi.
Peluso, N. L. and Lund, C. (2011). New frontiers of land control: Introduction. *Journal of Peasant Studies*, 38: 4, 667–681.
REGLAP (2011). *Rangeland Fragmentation in Traditional Grazing Areas and its Impact on Drought Resilience of Pastoral Communities: Lessons from Borana, Oromia and Harshin, Somali Regional States, Ethiopia* (by Fiona Flintan, Boku Tache and Abdurehman Eid).
Shimelles, T., Zahidul Islam, K. M. and Parviainen, T. (2009). *Effects of Land Tenure and Property Rights on Agricultural Productivity in Ethiopia, Namibia and Bangladesh*. Discussion Paper No. 33. Department of Economics and Management, University of Helsinki.
Smaller, C. and Mann, H. (2009). *A Thirst for Distant Lands: Foreign Investment in Agricultural Land and Water*. New York: IISD.
Time Magazine (2011). *Africa Blossoms: A Continent on the Verge of an Agricultural Revolution*. 31 October.
Watts, M. (1994). Development II: The privatization of everything. *Progress in Human Geography*, 18, 3: 371–384.
White B., Borras Jr, S., Hall, R., Scoones, I. and Wolford, W. (2012). The new enclosure: Critical perspectives corporate land deals. *The Journal of Peasant Studies*, 39, 30–34: 619-647.
Wondwosen, Michago Seide (2016). *The Nuer Pastoralists: Between Large Scale Agriculture and Villagization. The Case of Lare District, Gambella, Ethiopia*. Future Agricultural Consortium, FAC (unpublished).
Woodhouse, P. and Ganho, A. S. (2011). *Is Water the Hidden Agenda of Agricultural Land Acquisition in Sub-Saharan Africa?* Manchester: University of Manchester.
World Bank (2010). *Rising Global Interest in Farmland: Can it Yield Sustainable and Equitable Benefits?* Washington, DC: World Bank.

10 Lake Tana

Source of disputes or collaboration over the Blue Nile?

Mats Hårsmar, Emil Sandström and Atakilte Beyene

Introduction

The announcement by the late Ethiopian Prime Minister Meles Zenawi in early 2011 that the Great Ethiopian Renaissance Dam (GERD) should be constructed on the Blue Nile ignited severe tension and debates with Ethiopia's old adversary Egypt. The incoming Morsi government in Egypt was later recorded suggesting various belligerent acts to stop the dam from being built (Kimenyi and Mbaku, 2015). It was not until March 2015 that Ethiopia reached an agreement with its Blue Nile riparian neighbours Sudan and Egypt to at least jointly study the effects of the dam further.

However, despite all the rhetoric around the GERD, it has been argued that the dam brings benefits to all three riparian states, and creates a win–win situation between Egypt, Sudan and Ethiopia, if agreement is reached on the distribution of water during the period when the dam is filled up and during periods of drought. The benefits to Sudan and Egypt are considered to be multiple, such as silt reduction and prolonged life of downstream dams (e.g. Sudan's Roseires and Egypt's Aswan High Dam), less flooding problems and increased production capacity of downstream hydropower schemes due to increased regularity of water flows (Whittington *et al.*, 2014) and potentially increased electricity access for Sudan in particular.

Against this backdrop, the current developments of irrigation schemes along the Blue Nile (also known as the Abay, meaning father) and in the Lake Tana area have received less hydropolitical attention. Irrigation schemes are particularly contributing to water consumption, leading to reduced water flows downstream. Should the number and extent of irrigation schemes increase in upstream riparian countries, transforming the 'blue' water of the Blue Nile into 'green' water (water stored in the soil that evaporates or transpires through agricultural crops and plants), this would potentially be more problematic to Egypt and its access to Nile water than yet another hydroelectric dam. Hydropower utilizes the fall of the water, but does not necessarily reduce the amount of water in the river, except for the losses of water due to increased evaporation in the reservoir.[1]

The current irrigation projects in the Lake Tana area play the double role of being at the core of national development efforts, while also serving as pilot projects meant to be evaluated and scaled up. Lake Tana – the source of the Blue Nile – is one of the highest located bodies of water in the world and is the largest lake in Ethiopia, forming the main reservoir for the Blue Nile, which alone contributes 59 per cent of the water that flows to the main Nile River. What unfolds in the Ethiopian Blue Nile and in the area of Lake Tana is thus important for the understanding of broader hydropolitical ramifications in the Nile.

This chapter explores past and present hydropolitical ambitions along the Ethiopian Blue Nile with a focus on what the Ethiopian government has pronounced as a 'growth corridor' area around Lake Tana in the north-western part of the country. The first part of the chapter provides a historical review of Lake Tana's central role in the Nile Basin's hydropolitical history. In the second part we describe a number of ongoing and planned water-related projects in the Lake Tana area, which in part three are discussed in relation to Ethiopia's broader national development efforts and changing hydropolitical relations with its riparian countries. The chapter concludes that contemporary developments along the Nile, with the signing of the Cooperative Framework Agreement (CFA), a radically stronger Ethiopian economy, political turmoil in Egypt in combination with huge potentials for silt reduction in the Lake Tana area has opened up a leeway for Ethiopia to go ahead with its long-held ambitions to develop the Blue Nile waters for irrigation and hydropower.

The hydropolitical history of Lake Tana

Large-scale hydropolitical ambitions are far from being something new to the Lake Tana area. On the contrary, between 1902 and 1956, the Lake Tana area was at the centre of several countries' hydropolitical ambitions where Ethiopia, Italy, Egypt, Sudan and, above all, Britain struggled to get control of the lake. The lake was considered a key element in Britain's hydrological planning and colonial ambitions to secure water that could be used for irrigation downstream. Already in the 1880s Britain had taken a serious interest in Lake Tana, and, in 1902, Britain and Emperor Menelik II signed the Anglo-Ethiopian treaty, where article III states that:

> His Majesty the Emperor Menelek II, King of Kings of Ethiopia, engages himself towards the Government of His Britanic Majesty not to construct, or allow to be constructed, any work across the Blue Nile, Lake Tsana, or the Sobat which would arrest the flow of their waters into the Nile except in agreement with His Britannic Majesty's Government and the Government of the Soudan.[2]

In 1906, ideas for turning the lake into a great water reservoir were proposed; but it was not until the 1920s that Britain started to set their plan in motion.

The initial British plan for damming the lake was received with resentment in Ethiopia, had support in Sudan (which was a British colony at the time) and opposition in Egypt (which became independent in 1922). Egypt feared that the dam would destroy the important virtual fertilizer (silt) flowing down nurturing the fields in the Egyptian Nile Valley, whereas Sudan hoped that an increased and more regular water flow would increase the yields and the area under irrigation. Initially, Britain hoped to secure an agreement with the Ethiopian Emperor Ras Tafari (later Haile Selassie) through bribes and through different kinds of leasing arrangements[3] (Tvedt, 2004: 119–120). Gradually, however, it became clear that Ras Tafari was not going to accept the deal since he instead was interested in a trade where he would be granted the Italian colony of Eritrea in exchange for the lake. In 1923, Britain agreed that Egypt would be offered no more than half of the Tana waters in return for proportional costs of constructing the dam and the abandonment of 300,000 feddans on the development of the Gezira cotton scheme at the border between Sudan and Egypt. When water demands in Sudan increased, Sudan would take over the full use of the Tana reservoir, in repayment to Egypt for the sum that Egypt had contributed for the construction of the reservoir (Tvedt, 2004: 122).

However, the Italians became aware of the plans and, in 1922, Italy protested against the 'secret' British plans, claiming that the Lake Tana area was a zone reserved for Italy (according to Article 4 of the 1906 agreement), which according to Italy's interpretation provided Italy with territorial rights and Britain hydrological rights. Britain responded by recalling the 1902 Anglo-Ethiopian agreement that stipulated that Tana was an issue to be negotiated between Ethiopia and Britain (Tvedt, 2004: 122).

In 1925, secret notes were exchanged between British Prime Minister Stanley Baldwin and Italy's Benito Mussolini on the Tana issue. This eventually ended up in an informal agreement where Britain provided exclusive rights to Italy's ambition to connect Eritrea with the Italian Somaliland with a railroad and Britain was in return given the right to build a barrier on Lake Tana. When the secret agreement became public, Ras Tafari proposed building a dam with Ethiopian funds, where the water could be rented to Sudan and Egypt. Britain's alliance with Ethiopia's historic enemy, beaten at the battle at Adwa and still occupying Eritrea, increased opposition in Ethiopia to sign any deal with Britain over the lake. Complicating matters further, in 1927 the USA emerged in the Lake Tana affairs. The Ethiopian government invited an American company (American J. G. White Engineering Corporation) to study the possibility of constructing a dam at Lake Tana.[4]

In October 1935, Italy invaded Abyssinia from Eritrea and bombed Adwa (a symbolic act of revenge from the Italian defeat in 1896). The bombardment turned world opinions against Italy and Haile Selassie became the man of the year in *Time* magazine. However, others, such as France, were interested in the Abyssinian question. A meeting was held between the French and British foreign ministers Pierre de Laval and Samuel Hoare to discuss various ways to bring an end to the war. They agreed that Mussolini could be given a mandate to rule all of Abyssina except the Amhara region. Ethiopia should be given the port of Assab on condition that they did not construct a railway, which otherwise would compete with the French railway from Djibouti. Britain, for its part, would take control of Lake Tana and the Nile watershed in the country (Tvedt, 2004: 173). When the so called Hoare–De Laval pact became public, it was met with public outcry, which forced British Foreign Minister Hoare to resign and the legitimacy of the League of Nations to dwindle. Despite this setback, the British Lake Tana policy continued. However, in 1937, stories in British newspapers claimed that Italy itself planned to construct a tunnel from the upper Blue Nile Basin to the dry lands in Ethiopia. In 1938, a new agreement between Italy and Britain was reached, reaffirming earlier agreements. But tension continued, and when Italy joined Germany in World War II in June 1940, Lake Tana affairs took yet a new turn. In May 1941, British and Ethiopian troops defeated the Italian occupiers and the British focus on Tana Lake was reinforced. New estimations were made to the effect that half the share of the lake reservoir would increase the Gezira irrigation scheme in Sudan by 200 per cent over a 30–40-year period (Tvedt, 2004: 178).

New proposals regarding the lake were now forwarded by Sudan, Egypt and Britain. Sudan suggested that a concession to draw water from the Lake Tana Dam should be given to Sudan, whereas Egypt proposed that the dam should be under Egyptian control and that the Lake Tana area could be exchanged with territory in Sudan. Britain suggested that a concession should be given freely to Britain by the newly installed government in Ethiopia as a 'gratitude' for the costs of the war with Italy (Tvedt, 2004: 178–181).

In 1946, the British published a new master plan for the Nile waters, *The Century Storage Scheme*, which among other things suggested the construction of dams along the river, including dams in Northern Sudan, the Central African lakes and the Lake Tana area, as well as the old plan of digging a canal across the Sudd (the Jonglei Canal). In the publication, Lake Tana was considered a masterpiece in Britain's hydropolitical planning, although the plan was not that much different from previous plans. The exception was that the plan this time was even more ambitious. It was suggested that the Lake Tana Dam should store even more water. Parallel to the publication of *The Century Storage Scheme*, the British also initiated yet another study of the Nile in Ethiopia, a report of conditions for developing hydropower in the Lake Tana area. This classified study suggested among other things a hydropower station of 58 megawatts (MW) at the Tisisat Falls and the construction of a tunnel from Lake Tana to the River Beles where it was estimated that the hydropower station could generate as much as 240MW (a considerable amount at this time). Representatives from Britain, Sudan and Egypt continued to detail the plans for turning the lake into a major water reserve in which the British hoped to transform the Nile irrigation system to serve its cotton interests in Sudan and Egypt.

After World War II, Ethiopia under Haile Selassie was well aware of the geopolitical role his country played as the water tower of the Nile. Haile Selassie was so obsessed by the Tana water that he even considered moving the capital from Addis Ababa to Bahir Dar at the shore of Lake Tana. Haile Selassie opted either to build the dam himself and sell the water to Sudan and Egypt, or to use the water for hydropower. Another option was to trade the dam for Eritrean territory, an idea that in 1949 reappeared in connection with Britain's and Egypt's pushing ahead with the Lake Tana plans. Ethiopia wanted support to claim back Eritrea. Britain was willing to accept, but Egypt wanted both the dam and restoral of its own territorial claims to the Massawa area in Eritrea.

The British left the Tana affairs just prior to Sudan's independence in 1956, only for a new potential hegemon to appear; Egypt made attempts to incorporate Sudan under its Unity of the Nile Valley programme and also made substantial efforts to include Eritrea under Sudan (Arsano, 2007: 88). However, when this turned out to be a failing strategy, Egypt continued with its plan to construct the Aswan High Dam with the purpose of controlling the Nile River water within its own borders. It was partly against this backdrop that Sudan and Egypt also signed the 1959 Nile Water Treaty, which gave Egypt and the Sudan 'historical rights' to use the Nile River water. Of the annual average of 84bcm (billion cubic metres) as measured at the Aswan High Dam, Egypt was given the right to use 55.5bcm and Sudan 18.5bcm, leaving 10bcm for evaporation and seepage. Underlying the exclusion of Ethiopia from the 1959 Treaty was also the treaty from 1902 between Britain and Ethiopia, where the former Ethiopian Emperor Menilik II supposedly disclaimed any right to the Nile and agreed not to take any measures that would reduce the availability of the Nile's water to the Britannic Majesty's Government and the Government of Sudan, including the building of any dams.[5]

In 1957, Haile Selassie also decided to ally with the USA and launched a major study of the Blue Nile for irrigation and hydroelectric power (the Abay master plan). The investigation took five years to complete and attempted to eliminate the annual flood of the Blue Nile by blocking its water behind four dams with a hydro-electrical capacity three times that of the Aswan Dam. It was also suggested that water should be transferred from Lake Tana to the Beles River by a tunnel and then made available for large-scale irrigation (LSI) in the valley downstream. Five hydroelectric power stations were envisaged in the upper stream, and the study indicated possibilities for irrigated agriculture and a resettlement of farmers from the denuded and overcrowded highlands of northern Ethiopia (Arsano, 2007). If managed as designed, the amount of available water could actually increase for downstream riparians. It was suggested that the Blue Nile Dams would be managed in conjunction with the Roseires reservoir and free it from siltation. Water could be released in May without sustaining the huge evaporation loss at Aswan and Nasser Lake. However, Egypt would then not benefit from receiving additional water in years of high floods that instead would be stored in the Blue Nile reservoirs and it would reduce the hydropower capacity at Aswan. As compensation Egypt was to receive additional water for irrigation. Ethiopia could, however, also choose to withhold the water in times of low rainfall to threaten a disaster further downstream in the Nile Valley (Collins, 2000).

The proposal revived the Egyptian fear of losing its control over the precious Nile waters. In 1976, Egypt signed a military pact with Sudan, promising to protect Sudan from outside attacks and by posting some 50,000 soldiers in Sudan, as a sign of force against any Ethiopian attempts at hindering the flow of the Blue Nile (Arsano, 2007). When Ethiopia in 1977 reasserted its rights to the Nile waters at a UN water conference and few years later at an African ministerial meeting in 1980, Egypt started to press for a Nile water commission and continued to proceed with its own master plan to tap more water from the White Nile, which included the Jonglei canal across the Sudd (Collins, 2000; Oestigaard, 2012). Lake Tana's hydropolitical role became overshadowed by Egypt's own grandiose plans, only to resurface again some 60 years later by the launching of the GERD and the establishment of a number dams and irrigations schemes in the Lake Tana area.

At the end of the Cold War, Egypt initiated technical cooperation along the river basin. However, the hydrological monitoring undertaken by the HYDROMET, as well as the technical cooperation within the TECCONILE (Technical cooperation committee for the promotion of the development and environmental protection of the Nile River Basin), were seen by the upstream riparian countries as attempts by Egypt to gain increased control over the Nile River (Earle *et al.*, 2013: 121). The need for increased riparian trust was seen as essential by a number of different actors. A first step in that direction was taken in July 1993 when President Mubarak of Egypt and President Zenawi signed an agreement entitled Framework for General Co-operation between the Arab Republic of Egypt and Ethiopia. The parties agreed that the use of the Nile water should be dealt with by experts from both countries and 'on the basis of rules and principles of international law'. A section of the agreement specified that the parties have to consult and cooperate on projects that are mutually advantageous, such as various development schemes. This has been interpreted as binding Ethiopia to consult with Egypt before it starts to use the Blue Nile River's water for development purposes (Kimenyi and Mbaku, 2015: 103f).

Building on the experiences of the TECCONILE collaboration, the Nile Basin Initiative (NBI) was launched on 22 February 1999 by the Council of Ministers of Water Affairs of the Nile Basin Countries (Earle *et al.*, 2013: 12). The purpose was to shift the perspective from sharing of water to the sharing of benefits from the water, while simultaneously increasing mutual trust and collaboration. All the riparian countries participated in this initiative from the outset. The lead concept of the new initiative was integrated water resource management. The objective was to facilitate cooperation and 'to achieve socio-economic development through the equitable utilisation of, and benefit from, the common Nile Basin water resources' (NBI, 1999). To achieve this, the NBI implemented a set of joint programmes and organizational structures. Within one of these, the Eastern Nile Subsidiary Action Programme (ENSAP), an agreement was reached in 2004 with the planning of up to 100,000ha of irrigation, including 20,000–30,000ha of irrigation schemes that immediately could be implemented in Egypt, Sudan and Ethiopia along the Blue Nile. This agreement was later ratified in 2007 when actual projects were initiated with financial support from the World Bank and others. The agreement can be seen as milestone as regards the Blue Nile and Lake Tana hydropolitics. It was the first time the three riparian countries were in consent that water infrastructures could be built in the Lake Tana area.

Seen over a period of a decade, a process of trust building seemed to have emerged as a result of the political and practical collaboration within the NBI as well as earlier transboundary collaborative initiatives. Projects within the NBI framework were selected, shaped and sequenced in order to contribute to building trust. Such effects have also been noted (Earle *et al.*, 2013: 24, 29). Although the NBI generated important steps towards realizing a number of water-related projects, the riparian states could not settle the central issue of reaching an agreement on the 'equitable share' of the water. The historical exclusion of upstream riparian countries from the use of the Nile's water has been a constant concern and upstream riparians have pushed for a new judicial agreement that would take into consideration what they perceive as their legitimate rights to the water. As a result, the CFA was signed on 14 May 2010 in Uganda by four of the upstream riparian countries (Tanzania, Uganda, Ethiopia and Rwanda), with Kenya signing five days later and Burundi on 28 February 2011 (Salman, 2013). Sudan and Egypt refused to sign and, in protest, abstained from further collaboration in the NBI other than meetings at ministerial level (the Nile-COM). This also implied that their participation in and financing of the ENSAP came to an end, even though these projects were intended to run independently of the political process (Salman, 2013). In the following, we will describe what is unfolding in the Lake Tana area.

Contemporary developments in the Lake Tana area

Lake Tana (see Figure 10.1) is the largest lake in Ethiopia where the Blue Nile (Abay) originates. The lake basin is of national and regional importance and has been pointed out by the government as a development corridor for economic growth (MoFED, 2006). The population of the lake basin was estimated to be more than 3 million in 2007 (CSA, 2003) where the majority of the population pursue agriculture as their main source of livelihoods (CSA, 2003). Bahir Dar, located on the lakeshore, has more than 200,000 people and is the capital city of the Amhara Regional State. The Lake Tana Basin is rich in cultural and natural heritages (Heide, 2012) and, in 2015, it was designated as a protected biosphere area by the United Nations Educational, Scientific and Cultural Organization (UNESCO, 2015). Lake Tana is also renowned for being an important religious centre, and a vital a place in the cosmology of the influential Ethiopian Church; the lake's islands contain more than 20 churches and monasteries with deep historical roots (Oestigaard and Gedef, 2013).

The total potential irrigable land in Ethiopia is estimated to 3.5 million hectares, out of which 1.49 million hectares are situated in the Blue Nile Basin

(Awulachew *et al.*, 2007). The estimated irrigation potential along the Abay River itself is some 711,000ha (Arsano, 2007: 147, 157, Table 11). The Lake Tana Basin has four major river tributaries: Gilgel Abay, Ribb, Gumera and Megech. These rivers account for 93 per cent of the lake inflow (Kebede *et al.*, 2006). According to Wale *et al.* (2013), these tributaries have a total of 130,508ha of irrigable land (see Figure 10.1). This potentially irrigable area is

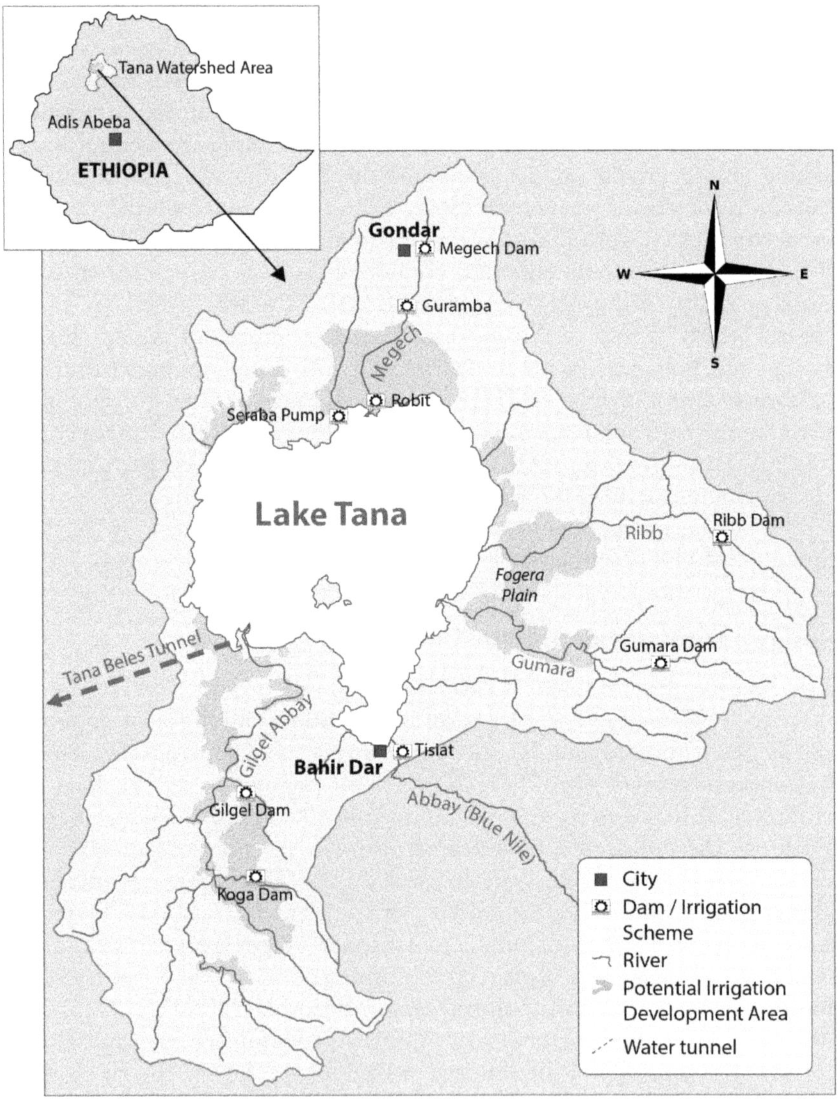

Figure 10.1 Map showing Lake Tana's water catchment area, with potential areas for irrigation and dams for irrigation, silt catchment or hydropower

Source: Henrik Alfredsson, Nordic Africa Institute.

distributed across a total of 106 medium-scale (200–3,000ha) and 11 LSI areas (greater than 3,000ha) (Wale *et al.*, 2013).

In Ethiopia's current master plan for the Blue Nile Basin, the government has identified 123 large- and medium-scale irrigation projects, which together cover an area of 526,000ha. They have also designated a supplementary area of 100–150,000ha for small-scale irrigation projects. In addition, five sites suitable for hydropower projects are identified (MoWE, 2014a). Hitherto, few of these water management interventions have materialized. However, a vast number of water management interventions are currently underway in the Lake Tana and the Blue Nile Basin water catchment areas (see Table 10.1). These interventions include hydropower, irrigation and drainage schemes, some of them in line with the earlier ENSAP collaboration.

Approximately 11 per cent of the basin is estimated to be suitable for surface irrigation, but less than 0.25 per cent of the basin could be irrigated consistently by the runoff river system (Wale *et al.*, 2013). This implies that the irrigation potential can only be met by increased dry season flows (Wale *et al.*, 2013). This is the main reason why construction of dams is seen as indispensable in order to store the rain (which mainly falls during three months of the year) and distribute it for energy and irrigation purposes during the dry season.

Table 10.1 Dams and irrigation schemes in the Lake Tana Basin area

Name of irrigation scheme/dam	*Irrigable area (ha)*	*Dam capacity (million m³)*	*Status*	*Number of rural households enrolled in the scheme*
Gilgel Abay	12,850	563	Feasibility study completed	NA
Gumara	14,000	60	Feasibility study completed	NA
Ribb	14,500	234	Under construction	11,500
Megech Dam	7,300	185	Under construction	13,500
Megech Seraba pump	4,000–5,250	Lake withdrawal	Under construction	2,000
Koga	7,000	81.3	Completed	10,000
Megech Robit	6,000–6,500	Lake withdrawal	NA	NA
Megech Guramba	6,530	Lake withdrawal	NA	NA

Source: SMEC (2008); Alemayehu *et al.* (2010); McCartney *et al.* (2010); MoWE (2014a).

Clear and exhaustive information on the actual number of dams and size of irrigated land in the Tana Basin area is not publicly available. Available reports hardly distinguish between projects that are planned, under construction or already operational. However, the Ministry of Water and Energy of Ethiopia has conducted feasibility and design studies of irrigation projects comprising a total area of 62,457ha in the Lake Tana Basin area (MoWE, 2014a). The same report also indicates an additional 14,000ha of planned irrigation registered under a different project name (Gumara Irrigation Scheme), which also is located within the Lake Tana Basin. Hence, according to government sources, the total size of planned irrigation area for the basin is 76,457ha. Irrigation projects that are operational, under construction and planned for are summarized in Table 10.1, and those that are operational or under construction are described further below. In addition to projects mentioned in the table that concern the Lake Tana Basin, an area of 75,000ha is already irrigated for sugar cane cultivation further down the River Beles, as well as a number of large-scale commercial irrigation projects within the wider Blue Nile water catchment area (c.f. Bossio *et al.*, 2012).

As mentioned, Ethiopia has anchored its decision to expand irrigation in the Lake Tana area with Sudan and Egypt through the Nile Basin framework of the ENSAP in an agreement settled in 2007. The international financing partners, notably the World Bank as initiator and coordinator of the Nile Basin Trust Fund, stick to this agreement as a necessary condition for their continued support of the Ethiopian irrigation projects. However, the political climate around the NBI is substantially different today, as compared to 2007. Sudan and Egypt no longer participate in any collaboration except for discussions at the ministerial level. Furthermore, Ethiopia is, through projects financed by other partners, particularly from its own financial sources, already at a level where 100,000ha are to be irrigated in the Blue Nile Basin, and almost double the amount is being planned for further irrigation expansion.

The Ribb Dam and the Fogera Irrigation Scheme

The Ribb Dam is located some 50km upstream the Ribb River on the east side of Lake Tana. The dam is situated between 1,872 and 1,945.5 metres above sea level (masl) and will have a storage capacity of about 234 million cubic metres of water encompassing a water surface area of about 1,020ha (full capacity) (MoWE, 2014b). The dam will be used to regulate the flow of water benefiting a separate irrigation scheme covering some 14,500ha located 30km downstream at Fogera Plain. Another major objective of the Ribb Dam is to collect and store silt from the Ribb River. Contrary to the other dams in the area, the Ribb Dam has no silt outlet and will thus serve as an end station for eroded soil. During the next 100 years, which is the calculated lifetime of the dam, the silt level is expected to rise up to a height of 1,910masl and serve as the 'dead storage level' of the dam. Construction of the Ribb Dam started in October 2008 and it was planned to be finished in June 2014. However, the state-owned contractor (Ethiopian Water Works Construction Enterprise) has faced severe difficulties in terms of fourfold cost increases, including exchange rate losses and lack of machinery and manpower. The revised plan was to finish the dam by June 2016. However, an assessment based on a site visit in September 2015 indicates that construction realistically may stretch out for additional years given the resources and pace of work observed.

The construction works of the canal and drainage system in the Fogera Plains commenced in late 2015. The Fogera Plain is normally flooded during the rainy season, often with severe negative effects on buildings and infrastructures. Despite this, the plain is densely populated, with some 60,000 residents involved in a mixed farming system with a high dependency on livestock breeding. Smallholder rice cultivation was first introduced and dispersed in the 1970s and 1980s during the Communist Dergue regime period (Gebeyehu and Genet, 2015). During the rainy season, rice is cultivated and is increasingly becoming the main source of income for farmers (Gebeyehu and Genet, 2015). The construction of the dam and irrigation scheme is expected to bring major improvements in local livelihoods with a more regulated water flow during the rainy season and increased water availability during the dry season.

The Megech Dam and irrigation projects

The Megech Dam is located at the northern side of Lake Tana along the Megech River. The objective of the dam is to store 185 million cubic metres of water that will irrigate the Dembia Plains. The Megech Dam is similar to the Ribb Dam in that it will serve as storage for silt, as well as contribute to flooding regulation and drainage. The Megech Dam is also anticipated to provide drinking water for the inhabitants in Gondor. The planned irrigation area of the Megech Dam is some 7,300ha. Similarly to the Ribb Dam, this dam is being built by the state contracting enterprise, and financed by the Ethiopian government and bilateral partners.

Another irrigation scheme is also under construction in the Megech River Basin, the so-called Megech Pump Irrigation and Drainage Scheme (MPIDP). The MPIDP, also known as the Seraba irrigation scheme, is located in Dembia Woreda, at the northern edge of Lake Tana between the Megech River and the port of Gorgora. The plan is to irrigate 4,000–5,000ha of the Dembia floodplain by water pumped from the lake into a canal and then distributed by gravity to approximately 2,000 farming households (SMEC, 2008). As distinguished from the combined dam constructions and irrigation schemes, the Megech Seraba scheme is irrigated through pumps lifting water out of the lake by gravity for distribution to the cultivated fields.

The Koga Dam and Canal Irrigation Scheme (KDCIS)

The KDCIS is the only large-scale scheme that is fully operational in the Lake Tana Basin. The dam was constructed in 2008 by the Ministry of Water and Energy of Ethiopia and the irrigation scheme was partly financed by the African Development Bank and was inaugurated in 2010. The KDCIS is located 35km south of the regional capital Bahir Dar, in Mecha District. The KDICS draws water from the Koga River, a tributary of the Gilgel Abay River. The Koga River Basin covers an area of 266km^2 and it stretches as high as 3,200m where the 64km long Koga River has its source (Gebreyohannis *et al.*, 2009). The water reservoir of the Koga dam has a storage capacity of 81.3 million cubic metres of water with a surface area of about 1,750ha. The irrigation scheme comprises a total surface area of approximately 7,000ha involving about 10,000 rural households.

The canal system of the KDCIS stretches from the main dam outlet all the way to the individual irrigation plots. The irrigation system is dependent on gravitational forces and the canals are aligned along the contours of the topography. This extensive network of canal systems comprise a 19.7km long concrete primary canal, along with 11 smaller water storage dams; 42.4km of concrete secondary canal; 112km long tertiary canal; and a 110km long drainage canal system. The KDCIS has also a catchment area of 22,000ha of land. The hydrological pattern of the Koga River varies widely during the year. About 70 per cent of the water flow in the Koga River occurs during the main rainy season between July and September, and the lowest flow occurs between December and March (Gebreyohannis *et al.*, 2009). Hence, during the rainy season, farmers use rainfall to cultivate their crops while the dam's outlet is closed in order to fill the dam with water. The water is released only during the dry season.

The building of the KDCIS implied displacement and resettlement of people and redistribution of farms. Households that were living and farming in the reservoir area were displaced (Eguavoen and Tesfai, 2012). According to local officials, displaced farmers were compensated for the properties they lost and many of them were also given land within the irrigation schemes. In the downstream case, complete redistribution of land was carried out. In this process, areas that fell into the designed irrigation topography were included in the redistribution process. Hence, communal and grazing land as well as arable land was included in the redistribution of lands.

According to the irrigation development policy of the country, management of irrigation schemes is considered a joint responsibility between involved stakeholders. For schemes like KDCIS, which targets smallholders as the primary beneficiaries, inclusion of smallholders is seen as critical by the government. There are legal procedures that guide how farmers should organize themselves into producer and service co-operatives. Directives that focus on irrigation co-operatives are being developed but not yet in place. These include cost recovery schemes, and the introduction of water fees to pay for the operation and maintenance of the canals and costs for upstream water shed management.[6]

The farmers in the scheme have formed 12 primary water co-operatives and one union for the whole scheme. Each co-operative is, in turn, subdivided into water blocks and water units that have been tasked to regulate water allocation, keep track of water use rotations and mobilize farmers to carry out maintenance work of the water canals. Conflict resolution and conflict management are also carried out by the co-operatives. The Water Union mainly provides tasks related to delivery of agricultural inputs (fertilizers, pesticides and seeds) and storage facilities. The Water Union also helps take farmers to the market and assists in the multiplication of improved seeds.

In addition to tasks related to the functioning of the irrigations scheme, management of the catchment is also important for the sustainability of the scheme. The size of the catchment area is about 22,000ha and land use in the upstream area is likely to impact upon the dam and the irrigation scheme. The original plan was to reduce soil loss by 50 per cent through watershed management programmes such as soil and water conservation and forestry development. Apart from this plan, concrete plans on how to mobilize people, how to cover the expenses for conservation practices on a continuous basis and how to address growing population pressure in such areas are largely missing.

As a plan, there is an optimistic expectation that downstream beneficiaries will be able to cover the costs required to maintain the environmental services incurred by the people in the upstream areas. This, however, is unlikely to happen in the near future. First, the profitability and productivity of the irrigation agriculture has not been evaluated yet. Poor access to high-market values, poor post-harvest and limited agro-processing facilities in the area are major bottlenecks. While large-scale schemes are seen as a key for raising smallholder farmers' productivity, probably even necessary, the inclusion of a vast array of different smallholders may also become difficult due the high transaction costs involved in organizing the numerous famers involved.

The Tana-Beles Hydropower and Irrigation Scheme

Related to Lake Tana is also the Tana-Beles hydropower project. This project involves the transfer of water from the western side of Lake Tana to the Beles River via a 12km long tunnel (McCartney *et al.*, 2010). The Beles River emanates from the hills on the north-western side of Lake Tana and flows due west in the opposite direction from the Lake Tana Basin. It then runs west until it meets the Blue Nile (Abay) just before the border to Sudan. Hence, it forms part of the Abay Basin even though its natural flow is not connected to Lake Tana. The Beles River was identified for its hydroelectrical potential as early as the beginning of the twentieth century.[7] The idea of transferring water from

Lake Tana to the Beles River was first presented in the Abay Basin master plan, which was developed in 1958–1964. Four hydroelectric power stations as well as irrigation schemes were proposed in the original master plan for the Beles River (Collins, 1990: 277).

The construction of the Tana-Beles scheme was initiated following the Ethiopian drought of 1984–1985. The area designated for irrigated agriculture was 220,000ha. Resettlements occurred both in the areas directly affected by the schemes, but also through immigration into the area from drought-stricken areas in the North (Wollo, Tigray, Gondar and Gojam) (Tesfaye, 2001: 49). The 1984/1985 project was never aimed at serving the existing local population, but rather to involve settlers from outside. Many of the original inhabitants belonging to the Begga ethnicity left the area. Arsano (2007) claims that even though information about the project was classified, it was already in the late 1980s, possible to conclude that the project largely was a failure. Political resistance and instability together with poor management were the main explanations behind the failure. A new initiative was taken during the 1990s, but it was not until 2006 that the Ethiopian government awarded an Italian company the contract to build the Lake Tana-Beles project. The power plant receives water from the lake to produce electricity; the water is then discharged into the Beles River. The hydropower plant has an installed capacity of 460MW. It is also expected to provide water for the irrigation of 140,000ha.

The Tana-Beles hydro project was constructed under great secrecy and military protection and was inaugurated in connection to the completion of the irrigation scheme in May 2010, at the same time as the CFA was signed. Egypt was simultaneously highly critical of the establishment of the Tana-Beles project. It may, however, not represent such a big threat to Egypt, since the hydropower construction and the transfer tunnel from Lake Tana may not decrease the flow of water in the Blue Nile downstream. Nevertheless, the diversion of water immediately below the hydropower scheme into an irrigation scheme currently feeding a 75,000ha sugar cane plantation may represent a threat to Egypt's access to water, particularly when considering that sugar cane is a highly water consuming crop.

Net water withdrawals and silt as a lubricant for downstream acceptance

The net water withdrawal from the irrigation systems is considered to be substantial. Alemayehu *et al.* (2010) have estimated that the net water withdrawal demand of the anticipated irrigations projects to be in the range of 563 to 676 million cubic metres per year, causing about a 0.4 metre drop in lake depth (excluding the Tana-Beles irrigation scheme). The estimated figure on net water demand, however, does not necessarily translate into a similar loss of water flowing downstream, given the complex hydrology of the Nile. Nevertheless, these indicative figures show that there are potentially non-trivial water use impacts, particularly for Tana Lake itself. The cumulative effects of the water infrastructure being put in place in the Tana area, with increased silt reduction, may, to some extent, counterbalance some of the negative impact from the expected substantial water loss from the irrigation schemes.

The shallow nature of Tana Lake means that there is an increasing risk of sediment pile up as a result of intensive soil erosion in the catchment (Poppe *et al.*, 2015). Reduced silt flow can thus be seen as a necessary intervention to save Lake Tana from slow disappearance, but also as a lubricant for downstream acceptance, considering that the dams currently being put in place implicitly also reduce the amount of silt that flows into the forthcoming GERD, as well as into the dams in Sudan and Egypt. In this sense, the construction of the dams in the Lake Tana area may provide important services to downstream riparians. By reducing the amount of silt flowing downstream, as well as increasing the regulation of the water flow over the year cycle, the benefits for the downstream riparian states may override the possible negative aspects of significantly increased irrigation upstream.

The emergence of new state–water–farmer relations

The irrigation schemes currently built and planned in the Lake Tana region follow a joint model of combining LSI infrastructure with smallholder production. Altogether, the Ethiopian government appears to provide a systematic approach in terms of putting in place a system of dams and irrigation schemes in the Lake Tana area. To a large extent, the present attempts to tame the Nile in the Lake Tana area (and probably elsewhere in Ethiopia) can be seen as a form of state-initiated social engineering, similar to what Scott (1998) has described in his nominal work 'Seeing like a state'. The irrigation systems in the Tana area share the characteristics similar to what Scott has described as:

1 an administrative ordering of nature and society;
2 the dominance of a 'high-modernist' ideology;
3 an authoritarian state able and willing to use its full weight of coercive power to put high-modernist designs into being; and
4 the absence of a civil society capable of resisting such plans.

(Scott, 1998: 4ff)

Comparable to what Scott has described, the Ethiopian approach to tame the Nile in the Lake Tana region is based on a modernistic developmental discourse that seems willing to justify 'short-term' sacrifices, through, for instance, displacement of people from their homes. The current plans and infrastructure development in the Lake Tana area are also, to a large extent, driven in an authoritative top-down fashion, while at the same time efforts are being made to engage its users to take responsibility over the infrastructure put in place. The Koga irrigation scheme, for instance, with its involvement of 10,000 rural households, includes various mechanisms to transfer the irrigation management, to the households, through the creation of a number of different forms of farmer-led organizations.

While the creation of new organizations, such as water co-operatives, etc., adapt farmers to new physical and social realities, they also tend to extend the realm of the state in domains of formalized decision-making over water allocation into everyday interactions of the people (c.f. Cleaver, 2012). This brings up an array of issues that relate to the role of the state, power inequalities and control over resources. The combined Koga Dam and irrigation scheme as an example involves a plurality of arrangements and different stakeholders. Far from being a simple site that increases farmers' access to water, the irrigation scheme is a site with multiple objectives. While some agencies are tasked to develop the infrastructure, others are responsible for the distribution of water resources. Yet others are involved in promoting new agricultural practices and inputs, etc. In this process, farmers are increasingly subsumed into the objectives of powerful actors, where they increasingly become mutually dependent on the state and the Nile water and, therefore, deliberatively or not, also indirectly in the government's role in international hydropolitics.

The point is not to claim that the Ethiopian government intentionally uses the water infrastructure to control its people and as a leverage in international hydropolitics. Rather, the build-up of large-scale water infrastructures in combination with enrolling a vast number of smallholders in tapping water directly from the Nile inherently increases the involvement of various state authorities in water allocation and distribution beyond single irrigation schemes. Hence, the tendency is that smallholders' dependency on the state tends to proliferate with the establishment of the large-scale schemes. This may in turn lead to a reinforced role of the state in making decentralized production and management activities work. In this recentralization process, increased state engagement becomes important in Ethiopia's water decision-making processes not just at the level of individual irrigation schemes, but also at the level of broader Nile Basin hydropolitics.

Lake Tana in a changing hydropolitical landscape

The contemporary irrigation developments in the Lake Tana area need to be seen in relation to both past and present hydropolitical trajectories. From the historical review, it is possible to observe that ideas behind the hydraulic infrastructures presently being put in place in the Lake Tana area are far from new. They were preceded by a long trajectory of failed attempts and plans to tame the water-catchment system of Lake Tana. The present attempts can be seen as a form of reinvention of ideas that where pushed forward by the British already at the beginning of the twentieth century and later refined during the Haile Selassie era and during more recent collaborative attempts under the NBI. However, only now has Ethiopia been able to move ahead with the plans in the Tana area and the Blue Nile water catchment area.

Ethiopia on the move

When the late Ethiopian President Meles Zenawi was interviewed on Egyptian television on the 19 July 2010, he gave a very direct answer regarding why his government pushed ahead with the GERD and other water-related investments despite protests from neighbouring countries: 'Ethiopia has reached a stage where it can build its own dams with its own money', he said.

Ethiopia's rate of economic growth is one of the five fastest growing economies in the world. Between 2006 and 2014, the Ethiopian economy experienced on average 10.6 per cent annual real gross domestic product (GDP) growth. From 2012 onwards, the growth rate has been somewhat reduced to 8.5 per cent in 2015 (6 per cent real per capita GDP growth). Projections for 2016 are a growth rate of 8.7 per cent (6.2 per cent real per capita GDP growth) (AfDB *et al.*, 2015). Poverty has also steadily decreased during the past 15 years. In 1999 to 2000, the incidence of poverty at the national level, and as measured by the national poverty line, was 44.2 per cent and slightly higher in rural areas. In 2011, this had been reduced to 29.6 per cent and the difference between rural and urban areas had decreased (AfDB *et al.*, 2010, 2015, Table 14). However, poverty still remains widespread in rural areas and many households pursue subsistence agriculture that is highly susceptible to natural hazards. The effect of the current El Nino has made about 20 million people (about 20 per cent of the population) dependent upon emergency food and shows that the agricultural sector is highly vulnerable and that many households easily fall into poverty.

Ethiopia's policies for poverty reduction are to a large extent linked to investments in the agricultural sector (IMF, 2011: 22, 45ff). In order to enhance further agricultural expansion, a combination of public investments in enabling infrastructure, such as irrigation schemes, and further attraction of foreign direct investments (FDIs) are important (IMF, 2011: 46f). In its Growth and Transformation Plan (GTP), the government points to agriculture as the major source of economic growth, where support to smallholder farming is seen as central (IMF, 2011). Increased applications of appropriate technologies are called for, such as soil fertility improvements and rapid expansion irrigation schemes for cultivation (IMF, 2011: 50). A host of other interventions are also foreseen, such as natural resource conservation, improved extension systems and livestock development; but expansion of irrigation together with various efforts to encourage farmers to shift from low- to high-value products in their cultivation is a central part of the plan (IMF, 2011).

Shifting hydropolitical landscapes

Given the rapid economic growth Ethiopia has experienced over a decade, and given its strategic role as a Western ally in the region, Ethiopia's relative position towards Sudan and Egypt has shifted. During recent years,

Ethiopia's neighbouring countries have undergone fundamental political change, with implications for Ethiopia's geopolitical position in Nile Basin politics. Following the 'Arab Spring', which brought down its long-time president, Hosni Mubarak, Egypt went through a transitory period with a Muslim Brotherhood government under Mohamed Morsi (June 2012 – July 2013), before former general Abdel Fattah el-Sisi, through a military coup and later through an election in May 2014, was elected as president. During this period, Egypt has mainly focused on its internal problems, and less on its relations to neighbouring countries.

Parallel to these events, Sudan was split up into Sudan and South Sudan after years of continuous conflict since the 1950s. In 2005, a peace agreement between the government in Khartoum and the southern SPLA rebel forces was signed. This agreement contained the important principle of South Sudan's right to self-determination and hence opened a referendum on the issue in January 2011. The referendum subsequently resulted in the founding of the new nation of South Sudan. However, a civil war broke out in late 2013, when President Salva Kiir sacked Prime Minister Riek Machar. The president accused his prime minister of having plotted a coup. Machar responded by organizing a rebellion in which his ethnic group, the Nuer, fought the president's ethnic group, the Dinka. Shortly after this, in early 2014, Uganda sent 1,600 troops into South Sudan in support of the government troops (Roozenbeek, 2014: 17ff). Ethiopia, which has its own Nuer minority in the south-western parts of the country, refrained from a military intervention. The Ethiopian government called instead for all foreign troops to be withdrawn from South Sudan. Other riparian countries have been less shy to take sides in the conflict. Sudan has also sided with President Salva Kiir, and so did Egypt.

Altogether these events have implied openings for a renewed Ethiopian role in the region, including changes in relations at a personal level between presidents, as well as the emergence of new trade and commercial relations between countries. Improved relations between Ethiopia and Sudan have, for example, been noted, where President Bashir al-Assad may freely come and visit a mosque in Northern Ethiopia without any threat of being handed over to the International Court in the Hague, despite international orders of arrest. Furthermore, two new roads are currently being built across the border between Sudan and Ethiopia to encourage trade and movement across the border.

The major factors we have discussed – a radically stronger Ethiopian economy and geopolitical shifts in the Blue Nile Basin – imply that Ethiopia today has received a leeway in relation to neighbouring countries. Ethiopian leaders have during many years declared that Ethiopia is free to use natural resources within its territory to the benefit of its own population. The Ethiopian government has much larger possibilities to put power behind those words today compared to the past, and it is against this background that we need to understand the current development of LSI schemes and hydropower constructions in the Blue Nile and the Lake Tana area.

Conclusion

The hydropolitical landscape of the Nile and in the Lake Tana area is today in many respects a very different landscape than what existed prior to the Lake Tana affairs in the past, although the human ambitions to 'tame' the water have remained. Over centuries, control over the Eastern Nile has been a balancing act, where Lake Tana was at the centre of international disputes. During the colonial period, the British tried to consolidate control over the basin, mainly to safeguard British cotton interests in Egypt. This policy was so consistent that London even 'supported' an Italian invasion of Ethiopia in 1935 in order to gain control over Lake Tana. After the British withdrew from Lake Tana affairs in the 1950s, Egypt essentially continued the British tradition of securing control over the water resources. Backed up with a strong position in the region, politically and military, Egypt has over the years opposed the establishment of hydropower and irrigation schemes in Ethiopia. For instance, the World Bank was during long periods hindered to support such investments due to the Egyptian position.

In 1999, the NBI introduced new dynamics, by focusing on sharing benefits from water use, rather than on sharing water. The NBI and the ENSAP created a space for Ethiopia to anchor its hydropolitical ambitions along the Blue Nile and the Lake Tana area, particularly with Sudan, which did not necessarily correspond with Egypt's interests, and thereby also contributed to 'breaking up' the long-held Egypt–Sudan dyad established through the 1959 Agreement (see also Chapter 5 in this volume). When Sudan and Egypt withdrew from the NBI collaboration after the signing of the CFA, Ethiopia continued to construct irrigation schemes that had been jointly agreed during the previous phase, without much quarrel with the neighbours.

Time has worked in favour of this Ethiopian approach, since political turmoil in downstream countries, together with a rapidly increasing Ethiopian economy, has shifted the geopolitical balances. The secession of South Sudan has turned Sudan's aspirations towards increased cross-border trade and cooperation with Ethiopia. Even if tensions remain, cross-border collaboration over Blue Nile River management has continued – and this time outside the NBI framework, in which the joint expert committee work on the GERD is one such obvious example.

Besides these rather recent geopolitical shifts, there has also been a shift in how Egypt and Sudan perceive the value of the 'virtual fertilizer', more popularly known as the Nile silt. Silt is no longer in much demand in the Nile Basin. It is rather seen as an obstacle for the longevity of the many hydropower constructions that have been put in place downstream in the river system during past decades. For Ethiopia, however, the silt has turned into a bargaining asset. The potential of silt reduction in the Tana area, in combination with the signing of the CFA, political turmoil in Egypt and a stronger Ethiopian economy, has opened a leeway for the country to go ahead with its long-held ambitions to use the water for irrigation and hydropower. Our assessment of the contemporary water investments in the Tana area also confirm that silt reduction has been one of the major aims of the irrigation schemes currently in place in the Tana area.

Considering the scale of the anticipated irrigation projects in the Tana and Blue Nile water catchment areas and elsewhere in Ethiopia, there are also new questions evolving both in relation to the increased water volume needed and emerging state–farmer–water relations that may create new kinds of water dynamics along the Nile. With increasing numbers of farmers depending directly on the Nile water for their survival, farmers are likely not just to withdraw more water from the Nile, they may as well become indirectly subsumed into Nile Basin politics, contributing to an additional layer of complexity in future Nile Basin hydropolitics.

Notes

1 Official Egyptian statements have earlier been made to the effect that Ethiopian development plans do pose a potential threat to Egypt (Kendie, 1999: 150f).

2 Treaties between the UK and Ethiopia, and between the UK, Italy, and Ethiopia, relative to the frontiers between the Soudan, Ethiopia, and Eritrea, May, 1902. Retrieved from http://treaties.fco.gov.uk/docs/pdf/1902/TS0016.pdf in January 2015. Present-day Ethiopia, however, rejects the 1902 Treaty, arguing that it was never ratified by Ethiopia and that the wordings differ in important respects between the English and Amharic versions of the treaty. Ethiopia also holds on to the Harmon Doctrine that a country is absolutely sovereign over the portion of an international watercourse within its borders, thus overriding earlier historical treaties.

3 A bribe of a few thousand pounds would be considered. However, Ras Tafari was more difficult to bring on board than the British had expected. Britain then raised the bid and suggested a leasing agreement for the lake, in which the government of Ethiopia would receive UK£50,000 on the signature of the lease and £150,000 upon completion and an additional yearly sum of £30,000 as long as the dam reservoir at the lake was in use (Tvedt, 2004: 119–120).

4 Involvement of an American firm on the Nile was one of the reasons why the British pushed for a Nile water agreement with Egypt in 1929 (Tvedt, 2004: 134; Kimenyi and Mbaku, 2015: 95.

5 See endnote 2.

6 Co-operative expert at the Water Department, personal communication, September 2015.

7 The potential of the Beles River was first identified by C. E. Dupuis as early as 1906 and further confirmed by R. P. Black in 1921, when the latter discovered its phenomenal potential for the generation of hydroelectric power if water could be transferred from Tana to the Beles gorge only a dozen kilometres away from the lake (Collins, 1990: 277).

References

AfDB, OECD and UNECA (2010). *African Economic Outlook 2010 – Public Resource Mobilization and Aid*. African Development Bank, OECD Development Centre and United Nations Economic Commission for Africa, Addis Ababa/Paris.

AfDB, OECD and UNDP (2015). *African Economic Outlook 2015 – Regional Development and Spatial Inclusion*. African Development Bank, OECD Development Centre, United Nations Development Program, Addis Ababa/Paris.

Alemayehu. T., McCartney, M. and Kebede, S. (2010). The water resource implications of planned development in the Lake Tana catchment, Ethiopia. *Ecohydrology of Water Ecosystems and Society in Ethiopia*, 10(2–4): 211–222.

Arsano, Y. (2007). *Ethiopia and the Nile*. PhD thesis, University of Zürich, 2004. Ref.: Andreas Wenger; Korref.: Kurt R. Spillmann.

Awulachew, S. B., Yilma, A. D., Loulseged, M., Loiskandl, W., Ayana, M. and Alamirew, T. (2007). *Water Resources and Irrigation Development in Ethiopia*. International Water Management Institute Working Paper 123. Colombo: IWMI.

Bossio, D., Erkossa, T., Dile, Y., McCartney, M., Killiches, F. and Hoff, H. (2012). Water implications of foreign direct investment in Ethiopia's agricultural sector. *Water Alternatives*, 5(2): 223–242.

Cleaver, F. (2012). *Development through Bricolage: Rethinking Institutions for Natural Resource Management*. London: Routledge.

Collins, R. O. (1990). *The Waters of the Nile: Hydropolitics and the Jonglei Canal, 1900–1988*. Oxford: Oxford University Press.

Collins, R. (2000). In search of the Nile waters, 1900–2000. In Erlich, E. and Gershoni, I (eds). *The Nile: Histories, Cultures, Myths*. London and Colorado: Lynne Reinner Publisher.

CSA (Central Statistics Authority) (2003). *Ethiopian Agricultural Sample Enumeration, 2001/2002 Results for Amhara Region, Statistical Reports on Area and Production of Crops, Part II B*. Addis Ababa, Ethiopia: Central Statistics Agency.

Earle, A., Nordin, K. A., Cascão, E., Rukundo, D., Seide, W. M. and Björklund, G. (2013). *Independent Evaluation of the Nile Basin Trust Fund – Final Report*. Stockholm: SIWI.

Eguavoen, I. and Tesfai, W. (2012). Social impact and impoverishment risks of the Koga irrigation scheme, Blue Nile Basin, Ethiopia, *Africa Focus*, 25(1): 39–60.

Gebeyehu, T. and Genet, M. (2015). State and agriculture in Tana area. Conference paper presented at workshop *Assessing Transitions and Transformations in Smallholder Agriculture in Ethiopia*. Bahir Dar University and Nordic Africa Institute, Bahir Dar, Ethiopia, 27–28 August.

Gebreyohannis, G. S., Taye, A. and Bishop, K. (2009). *Forest Cover and Stream Flow in a Headwater of the Blue Nile: Complementing Observational Data Analysis with Community Perception*. Stockholm: Royal Swedish Academy of Sciences.

Heide, F. (2012). *Feasibility Study for a Lake Tana Biosphere Reserve, Ethiopia*. Federal Agency for Nature Conservation, Germany. Available at www.bfn.de/0502_skripten.html (accessed 10 February 2016).

IMF (International Monetary Fund) (2011). *Federal Democratic Republic of Ethiopia: Poverty Reduction Strategy Paper: Growth and Transformation Plan 2010/11 – 2104/15*. Washington, DC.

Kebede, S., Travi, Y., Alemayehu, T. and Marc, V. (2006). Water balance of Lake Tana and its sensitivity to fluctuations in rainfall, Blue Nile Basin, Ethiopia. *Journal of Hydrology*, 316: 233–247.

Kendie, D. (1999). Egypt and the hydropolitics of the Blue Nile River. *North East African Studies*, 6(1–2): 141–169.

Kimenyi, M. and Mbaku, J. (2015). *Governing the Nile River Basin: The Search for a New Legal Regime*. Washington, DC: Brookings Institute.

McCartney, M., Alemayehu, T., Shiferaw, A. and Awulachew, S. (2010). *Evaluation of Current and Future Water Resources Development in the Lake Tana Basin, Ethiopia*. International Water Management Institute Research Report 134. Colombo: IWMI.

MoFED (Ministry of Finance and Economic Development of Ethiopia) (2006). *A Plan for Accelerated and Sustained Development to End Poverty (PASDEP) 2005/06 – 2009/10.* Volume I: Main Text. Ministry of Finance and Economic Development, September 2006, Addis Ababa.

MoWE (Ministry of Water and Energy of Ethiopia) (2014a). Available at www.mowr.gov.et/index.php?pagenum=4.2 . Updated 18 March 2014 (accessed February 2015).

MoWE (Ministry of Water and Energy of Ethiopia) (2014b). *Rib Dam Construction Project-Current Status, Water Works Design and Supervision Enterprise.* Ministry of Water and Energy, the Federal Democratic Republic of Ethiopia, June 2014.

NBI (1999). *Shared Vision Program*, Nile Basin Initiative, available at www.nilebasin.org (accessed 10 February 2016).

Oestigaard, T. (2012). Water scarcity and food security along the Nile: Politics, population increase and climate change. *Current African Issues*, No. 49. Uppsala: Nordic Africa Institute.

Oestigaard, T. and Gedef, A. (2013). *The Source of the Blue Nile: Water Rituals and Traditions in the Lake Tana Region.* Newcastle: Cambridge Scholars Publishing.

Poppe, L., Frankl, A., Admasu, T., Dessie, M., Adgo, E., Deckers, J. and Nyssen, J. (2015). Geomorphology of the Lake Tana basin, Ethiopia. In Nyssen, J., Adgo, E. and Dessie, M. (eds). *Excursion Guide of the Tropi Lakes. 2015 Mid-Conference Excursion.* Bahir Dar, Ethiopia: Bahir Dar University.

Roozenbeek, J. (2014). *A New Stalemate: The Influence of South Sudan's Independence on the Nile Basin's Water Politics*, MA thesis, Sustainable Development, Uppsala University, Sweden.

Salman, M. A. (2013). The Nile Basin Cooperative Framework Agreement: A peacefully unfolding African spring? *Water International*, 38(1): 17–29.

Scott, J. (1998). *Seeing Like a State: How Certain Schemes to Improve the Human Condition Have Failed.* New Haven and London: Yale University Press.

SMEC (2008). *Hydrological Study of the Tana-Beles Sub-Basins.* Draft main report. Addis Ababa: SMEC International.

Tesfaye, T. (2001). *The Nile Question: Hydropolitics, Legal Wrangling, Modus Vivendi and Perspectives.* Munster: LIT Verlag.

Tvedt, T. (2004). *The River Nile in the Age of the British. Political Ecology and the Quest for Economic Power.* London: I.B. Tauris.

UNESCO (2015). *Media Services.* Available at www.unesco.org/new/en/media-services/single-view/news/twenty_new_sites_added_to_unescos_world_network_of_biosphere_reserves/#.VrH90bIrKUl (accessed 10 February 2016).

Wale, A., Collick, A. S., Rossiter, D. G., Langan, S. and Steenhuis, T. S. (2013). 'Realistic assessment of irrigation potential in the Lake Tana basin, Ethiopia'. In Mekuria, W. (ed). *Rainwater Management for Resilient Livelihoods in Ethiopia: Proceedings of the Nile Basin Development Challenge Science Meeting*, Addis Ababa, 9–10 July 2013. NBDC Technical Report 5. Nairobi, Kenya: International Livestock Research Institute. Available at http://hdl.handle.net/10568/33929 (accessed 10 February 2016).

Whittington, D., Waterbury, J. and Jeuland, M. (2014). The Grand Renaissance Dam and prospects for cooperation on the Eastern Nile. *Water Policy*, 16(4): 595–608.

11 Water, national identities and hydropolitics in Egypt and Ethiopia

Terje Oestigaard

Introduction

Individual identities and national ideologies structured around water are intrinsic not only to religions in the Nile Basin but also to unfolding politics, although often in subtle ways. The very identities structured around water work at different levels – individual, national and regional in between states – sometimes overlapping but often contradicting, causing unity but also tension. Dams, large-scale land acquisitions (LSLAs) and irrigation schemes consuming water may cause domestic conflicts, but also conflicts between states in a transboundary river basin. A central element in this constitutive process of collective identities in national ideologies is *control* and *security*, which are two fundamental aspects of autonomous state governance. Control of water is a powerful tool, which state leaders have used to mobilize collective identities of nations as well as shaping perceptions of other countries. A national ideology ideally unites all people in a country. According to Benedict Anderson, the nation can be seen as an imagined community or representing a deep, horizontal comradeship regardless of the inequality and exploitation that may prevail, and this makes it possible for so many millions of people, not so much to kill, as willingly to die for the nation or such imaginings (Anderson, 1993: 7). Water is not only the basis for all life, but also worth fighting for, and scenarios of water wars, particular in between Egypt and Ethiopia, have frequently been portrayed in the media fuelled by political rhetoric.

Hence, the aim of this chapter is to address different and changing ideologies and collective identities shaping part of the unfolding hydropolitics. Politics, negotiations and agreements of water allocations are obviously about different needs and water as a physical resource, but it is more than just mere shares of water. Why certain arguments in discourses are presented as justifiable reasons legitimizing rights to water is founded in ideologies. In this context, rather than perceiving ideology in a Marxist sense as an illusion concealing means of exploitation, I will use a more collective and inclusive definition of ideology as 'the set of beliefs characteristic of a social group or individual' relating directly to individual and national identities. Identities are constitutive in national ideologies in various but often contradicting ways; but collective identities tend to be presented and understood more coherently and reductionist when describing one group or nation in relation to another.

The Nile has throughout history linked Egypt and Ethiopia together. The long historical trajectories of collective identities structured around the Nile continue to shape national ideologies and hence hydropolitics in the region. In this chapter I will therefore focus on parts of Egypt's and Ethiopia's perceptions of water as a national identity in relation to each other. This will be done by first examining the more distant past and then the recent history, since conceptions of a shared identity, even shrouded in an unknown past, may have relevance without everyone in a culture or nation knowing all origins and mythologies shaping the different levels of collective identities.

Water and nationalism

Following Jenkins, 'identity can *only* be understood as process, as "being" or "becoming". One's identity – one's identities, indeed, for who we are is always singular *and* plural – is never a final or settled matter' (Jenkins, 2004: 5), and one has several individual as well as collective identities. At a state level, nationalism is about constructions of identities, and it insists on the linkage of social organization and culture. 'Nationalism does indeed see itself as a universal, perennial and inherently – self-evidently – valid principle' (Gellner, 1997: 7). The nationalist principle requires that the political unit or territory and the 'ethnic' one to be congruent. Nationalism demands that within the political unit everyone is of the same culture, and the main function of culture is to 'reinforce, underwrite, and render visible and authoritative, the hierarchical status system of that social order' because that 'homogeneity of culture is the political bond, that mastery of a given high culture is the precondition of political, economic and social citizenship' (Gellner, 1997: 20, 29). Nationalism is an ideology, and *Oxford Dictionaries* defines ideology as 'a system of ideas and ideals, especially one which forms the basis of economic or political theory and policy' or 'the set of beliefs characteristic of a social group or individual',[1] the latter definition being close to perceptions of culture and as such also shaping collective actions. In practice, thus, there are fluid distinctions between nationalism, national ideology and collective national identities, and all of them relate in various ways to individual identities and perceptions of belonging to a larger community, but individual and collective identities are never identical. In the interplay between individual and collective identities, there is always negotiation, contradictions and ambiguities.

National identities can involve numerous events, processes, places and persons, and even responsibilities. On the one hand, the creation of collective identities often involves the past. The primary role of archaeology as a nationalistic discipline, for instance, has been to 'anchor' the nation by making it simultaneously timeless and very old, and therefore nationalism itself has its reasons and its roots in the past (Sørensen, 1996: 28; Oestigaard, 2007). This form of anchoring nationalism in the past is not restricted to archaeology, but also history as a discipline, including the recent history. Importantly, as has been emphasized, 'Nations exist … and the archaeologist's job … is to accept them as artefacts while constantly drawing attention to the process of fabrication' (Ascherson, 1996: viii), but that does not by definition invalidate an historic interpretation as such. The point is merely that there are, for example, some 3,000 years between the Egyptians' building the pyramids and the Arab conquest of Egypt and almost 1,500 additional years to today's Egyptians; contemporary identities are very different from the identities and people of the past. Still, nationalism makes the past an active resource for the present and national identities are justified and legitimized by reinventing the past, and, as a national ideology, an invented past may work better than a factual past.

Many of the national identities in Egypt and Ethiopia are structured around and rooted in the past and the ever-flowing water of the Nile. The Egyptian civilization has from the dawn of history and science attracted philosophers and scholars alike, not only because of its uniqueness in world history, but also because Egypt has, since antiquity, been attributed with two of the seven wonders of the ancient world and it is truly one of or perhaps *the* most remarkable civilization of the world. The saying 'Egypt is the gift of the Nile' is attributed to Herodotus and certainly has a deep resonance in the Egyptian culture and perception of the world. Ethiopia, too, has a glorious past, but emphasizing other aspects. Queen Sheba is a prominent national figure and her marriage with King Solomon resulted in their child, Menelik, who became the very founding father of the Ethiopian dynasties. Axum was also a prominent civilization and the monastic islands in Lake Tana are unique in the history of Christianity in Ethiopia, and not to mention Gish Abay – the legendary source of the Blue Nile – flowing directly from the paradise connecting heaven with earth and hence the water being utmost holy (Oestigaard and Gedef, 2013).

On the other hand, national ideologies are also about the present and the visions for the future. In many countries, national identification has often been justified by constructing Monumental Memories of the past (Brown, 1998: 79), but equally important, national ideologies have also emphasized 'ideas of progress and modernity' (Kampanje-Phiri, 2015: 582). Control of water and big monumental water infrastructures such as dams are crucial for the construction of national identities. The former prime minister of India has perhaps been most explicit. Not only did Nehru celebrate big *dams* as the 'temples of modern India', but when he laid the foundation stone for the Hirakud Dam in Orissa in 1948, he said: 'If you have to suffer, you should do so in the interest of the country' (op. cit. Skarstein, 2016: 150).

Control of water in the form of rain has also been instrumental to power. In traditional rainmaking ideologies, 'Without rain, people, plants and animals will die. In short, rain is life' (Sanders, 2002: 287). The main function of rulers of rainwater was to 'ensure the moral and material well-being of the population of that land and it [was especially] concerned with rainmaking or the control of floods, with the fertility of the soil for agriculture … normally controlled by a limited elite of priests and functionaries' (Ranger, 1973: 582). In Sudan, for instance, the kings were believed to punish people collectively with drought and other ill fortunes, while the people blamed the king for the disasters. Good rains and abundant harvests were seen as the work of the king (Simonse, 1992: 191). However, killing the rainmaker was the ultimate measure taken if disasters were prolonged, and a common rationale was: 'He is killing us so why should we not kill him?' (Simonse, 1992: 199). This principle, that the state's ultimate responsibility is to safeguard the security and welfare of its people, embodied and symbolized by the head of state, is as valid now as in the past – in the Nile Basin and elsewhere. Leaders incapable of providing these basic premises lose legitimacy and are in danger of being displaced, violently or not, as during the Arab Spring in Egypt. In today's modern world, the days of leaders providing and securing the wealth and health of their people by rainmaking is gone; but controlling rain and rivers by dams and large water infrastructures in agriculture and the energy sector become more important. If controlling water for the 'wealth of the nation' is important, it poses further challenges in transboundary river basins like the Nile.

Water and water laws represent a particular challenge to concepts of sovereignty. 'A sovereign state can be defined as an authority that is supreme in relation to all other authorities in the same territorial jurisdiction, and that it is independent of all foreign authorities' (Jackson, 2007: 10). However, there are no determining or necessary relations between international laws and geographical position, since flowing water disregards social and political invented borders, and the popular perception that upstream states always have a strategic advantage is not always the case, as with the Nile (for a thorough discussion of water law, see Tvedt *et al.*, 2015). The important aspect of water in relation to nation states and sovereignty is that the national signifier or identifier is shared by two or several nations. Although the waters of such a mighty river like the Nile take specific forms and shape in the due course of its flow, and significant mythologies and identities may be ascribed to any of these places and associated events, it is still the same water. A shared (tangible or intangible) resource like the water of a transboundary river for multiple national identities somehow contradicts theoretically the basic idea of national identifiers highlighting the uniqueness of a particular nation in relation to others. If it is shared, it is not unique for a nation and rather represents a parable that binds people together across other boundaries. Still, different people may ascribe different qualities and spin unique webs of significance and identities around the same waters.

Thus, the fluidity of water, always being universal and particular (Tvedt and Oestigaard, 2010), may open up new entrances for understanding the identification potential of water and how water and the Nile can be a shared resource and source of national identities, while at the same time cause conflicts.

Flows of identities from the past

In Egypt, the Nile is literally the life-giving artery of the country. Since time immemorial, the Nile has had a mystical and grand aura. According to Greek and Roman philosophers of antiquity, the Nile was the most spectacular river in the world: 'The Nile surpasses all the rivers of the inhabited world in the benefactions to humanity', Diodorus once proclaimed. Seneca, on the other hand, said that all rivers were '*vulgares aqua*', but the Nile was the 'most noble' of all watercourses. Arnobius described the Nile as 'the greatest of rivers' and according to Ammianus Marcellinus it was 'a river which is kindly to all' (Wild, 1981: 88). Two thousand years later, the Egyptian writer Jamal Himdan expressed it this way: 'The first civilization was the fruit of fortunate geographical marriage between Egypt and the Nile, hence if history was a father of Egyptians and Egypt was a Mother of the World the Nile is simply the great, great grandfather of human civilization' (Himdan, 1987: 787). Thus, the Nile has not only been fundamental to Egypt's economy and well-being, it has also been a central part of the Egyptian identity and changing religions, which has partly continued up to today.

From the ancient Egyptian dynasties onwards, the water of the Nile has been seen as a divine creation, albeit in different forms (see Figure 11.1), which also includes Christianity and Islam (Oestigaard, 2009, 2011). Among the Copts and the Desert Fathers in Egypt, Shenute was an abbot at the White Monastery from the latter part of the fourth century to the third quarter of the fifth century (Young, 1993: 17). Since the monastery was situated in a desert, water and the Nile played a prominent role. In *The Angel of the Waters*, Shenute wrote of the tense wait for the annual flooding of the Nile and described the annual inundation of the Nile as God's 'yearly mercy' (Bell, 1983: 107). Besa, who was his disciple and successor at the White Monastery, writes in his *The Life of Shenoute*:

> It happened one year that [the Nile] did not flood, and our father apa Shenoute knew from God the hidden reason for it. He also revealed the matter to the brothers with his tears flowing from his eyes, and said to us: 'Pray to God. I, too, will go into the desert and spend this week praying to the Lord. See that no-one at all comes to me.'
>
> (Bell, 1983: 72–73)

Shenoute told his disciples

> You know what I said to you that God has commanded that there should be no flooding of the land this year. Behold, then, I prayed to him, and he, as the good and merciful God, promised me that this year again he would cause the waters to come and cover the face of the land.
>
> (Bell, 1983: 73)

Water and the Nile were an important part of Christianity and 'Shenoute's self-proclaimed relationship with God gave him knowledge of what God required for salvation' (Krawiec, 2002: 56), and the coming of the annual flood was an important Christian event.

If the Christians in both Ethiopia and Egypt believed the river was imbued with divine qualities from heaven, so too did the Muslims of Egypt. In fact, the Muslims Islamized some of these beliefs and rituals because otherwise the life-giving source would have been in the hands of the Christians and their God. Maqrīzī's historical texts from fifteenth-century Cairo describe the Coptic Nile festivals. The first Arab Muslim governor of Egypt forbade the practice of

Figure 11.1 Identity makers in Egypt: the pyramids and Nile inundation

Source: Lehnert and Landrock, taken in 1924.

sacrificing a female virgin to the Nile, but due to this prohibition the Nile did not rise for three months. The governor began to despair and asked pious Caliph ʿUmar I for advice. The caliph replied that instead of sacrificing virgins, he should offer a piece of paper on which the caliph had written:

> Allah alone can cause the Nile water to flow. [The governor] threw the paper in the Nile, one day before the festival of al-Salīb, but in the mean-time, the people of Egypt were preparing to depart, because their welfare rested only on the Nile. However, on the day of the Salīb festival, Allah, the Almighty, caused the Nile to flow, reaching 16 cubits in one night, thus preventing harm from happening to the people of Egypt.
>
> (Lufti, 1998: 256)

Thus, the Egyptian Muslims Islamized beliefs about the Nile and saw it as a gift from Allah. This control of water was structured within religious domains, and as Tim Insoll has pointed out, religion is often the key building block of identity and 'can be conceived of as the superstructure into which all other aspects of life can be placed' (Insoll, 2004: 3), but it is not the only category. The origin and control of the water can be embedded in both secular and profane spheres, highlighting the multiple dimensions of water in identity constructions. This control has also to be seen in relation to Ethiopia's cultural heritage and identities structured by water and the Nile, in particular. Moreover, the fates of Egypt and Ethiopia have been closely linked throughout history. The Egyptian Abbas El-Tarabily wrote in 2002:

> I am just back from Ethiopia, the source of [84 per cent] of Nile water, which created Egypt. In fact, Egypt was born in Ethiopia and from the red mountains of Ethiopia came the water, which is the stuff of our life and the silt that created the fertility of Egyptian soil … I was telling myself that it is from this place that the journey of the great river starts and carves the red rocks of Ethiopia leaving thousands of canyons and rifts that seemed to be the blood veins of the Egyptian body.
>
> (Op. cit. Arsano, 2007: 73)

Thus, turning from cultural and religious concepts of water in Egypt to Ethiopia, in particularly in the Abyssinian highlands, there are specific water cosmologies that not only have shaped individual identities but also national and religious perceptions of Ethiopia. The source of the Blue Nile is generally seen as Gish Abay in Abyssinia or the outlet of Lake Tana in Bahir Dar (see Figure 11.2).[2] In the Ethiopian Orthodox Church, there are two particular religious stories with regards to the holiness of the Nile on Earth: one relating to Gihon and the other to Abay and Gish Abay. Following the Bible, the four divine rivers in heaven have different qualities and characteristics; one consists of honey, another of oil, the third of wine and Gihon consists of milk. The source of the Blue Nile or Gish Abay is believed to have its source in heaven. It is the River Gihon flowing from paradise linking the divine realms with the earthly sphere. The other story is about a particular holy man called Abune

Figure 11.2 Heavenly source, or Gish Abay in Ethiopia

Source: Terje Oestigaard.

Zerabruk at the end of the eighth century AD. According to the legend of the Ethiopian Orthodox Church, the Abune received from God the power to cure people for their physical diseases and spiritual sins by using water. Before he used the water for healing, he had to bless the stream (Gish Abay) in the name of the Father, the Son and the Holy Spirit. People using this water would also gain eternal life. Even people who were not yet baptised could be considered as being baptised from that time onwards since they had used the water from Gish Abay, or Gihon, which the river then was called.

Due to the growing popularity with regards to Zerabruk's religious deeds and powers, the Abune eventually faced difficulties and obstacles with regional rulers. Prior to his stay around Abay/Gihon where he prayed, he was forced to leave the place Tsima because of political resistance against his religious teaching. At Gish Abay, he was eventually arrested by the king's soldiers. Before he was arrested, the Abune stood at the source of Gihon River, one of the main waters or rivers believed to be the source of heaven, prayed and finally gave his seven sacred books to the source of the river. It is also claimed that God told him to give the books to the river. He was imprisoned for five years during which time some destructive forces or events were said to have happened in the jail; as a result, the king released the Abune. After five years in prison and two years of teaching and preaching at different places, the Abune finally came back to the source of Abay where he had kept his books. The Abune prayed thoroughly and was devoted to God and asked the river to bring forth the books he had given in custody while he was taken to prison.

Gish means 'bring forth', and when Zerabruk stood at the source of the Nile where he had deposited the seven sacred book seven years earlier, he said *gish* – bring forth. Then the river brought forth the books to the surface, which were undamaged after being hidden in the water. Surprised and pleased of this miracle, the Abune asked his old disciple Aba Zerufael to look at the miracle. He said 'Aba Eyi' in Amharic, which means 'Father, look', directing the disciple's attention to the fact that the river brought the books to the surface without being damaged. 'Aba Eyi' became *Abay*, which is the Ethiopian name for the Blue Nile. Hence, the name of the holy river and in particular the source – Gish Abay – refers to this religious event. After this miracle, the river changed its name from Gihon to Abay. And divine waters are not like other types of water, with implications for politics and identities (see Oestigaard and Gedef, 2013). While the story of Zerabruk is very important in creating personal identities in particular for people living close to the source, the stories about him are not well known in Ethiopia at large. This points to another aspect of the relation between individual and collective identities, namely that the totality of a place such as Gish Abay might be seen as particularly holy without everyone sharing or knowing all the local myths in detail. At a national level, the water from the source is seen as utmost holy, which throughout history has shaped relations between Ethiopia and Egypt.

From the eleventh century, the question of Ethiopia's alleged ability to interrupt the Nile's flow became generally acknowledged in Egypt (Pankhurst,

2000: 25–38). For centuries, the Christians had kept alive the idea that the coming of the Nile waters is one of the miracles of the Virgin Mary (Six, 1999: 53). During the reign of the Fatimid Sultan al-Mustansir, around 1089–1090, the flood failed and did not reach Egypt. The Arab writer al-Makin reports that the sultan sent the Patriarch Michael of Alexandria to Ethiopia to ask the Ethiopians to restore the flood, which they accordingly did (Pankhurst, 2000: 26). During a war between the Christians and Muslims, it is believed that the Virgin appeared saying that God had given the Christian Emperor Dawit of Ethiopia (1380–1413) the wisdom to divert the Nile. Learning of this power, the Muslims became fearful about their harvests and declared they were not the enemies of the Christians. When Dawit heard this news, he praised the Virgin (Six, 1999: 57).

According to the legend, the Virgin Mary also told the Ethiopian King Dawit:

> O my beloved and beloved of my son Jesus Christ, and now I have asked my son on behalf of you that you will go and rescue my nation, the Christians, and thus He has granted and made even for you your way. Get up and go. And He will perform through your hand many miracles … And God gave him wisdom and he stopped the river [Nile], so it did not descend into the land of Egypt, because there are no rains in the land of the people of Egypt; unless the water of the [Nile], which flows from Ethiopia, reaches them, they do not plough, they do not sow seed and they do not get water at all … [King Dawit said] Was it not said once: To restrain the water is like beginning a war, but the will of God, the Lord of the Christians, may come about.
>
> (Six, 1999: 66)

Since the Medieval period, it was commonly accepted, not only by the Egyptians but also by European rulers, that the Ethiopians were the masters at controlling the Nile. The oldest Arab source referring to Ethiopia's alleged ability to control and divert the Nile is probably the *Kitab al-Jawahir al-buhur*, written by Ibn Wasif Shah al-Misri (d. 1202–1203). However, the belief that the Ethiopian emperor had the power to control the Nile was a double-edged sword. In 1522, during the reign of Lebna Dengel (1508–1540), a monk told the Venetian scholar Alessandro Zorzi that the emperor could take Nile water 'so that it did not reach Cairo', but that he refrained from doing so for fear that the Muslims 'would ruin the churches and the Christian monks who are in Jerusalem and those in Egypt of which there are many' (Pankhurst, 2000: 30). Moreover, the Ethiopian abun was appointed from Alexandria, and 'if Ethiopia was the source of the Nile for Egypt, Egypt was the source of the *abun*, the Egyptian metropolitan bishop, for Ethiopia' (Erlich, 2002: 9) and the abun was second only to the emperor. 'The Egyptian card was the sultan's or the Coptic patriarch's ability to delay or avoid sending an *abun*' (Erlich, 2002: 22). During the nineteenth century, the 'idea of diverting the Nile' was again seen as a miracle of the Virgin Mary, thereby reanimating a theme that had been a challenging one in the relationship between Egypt and Ethiopia since early times (Six, 1999: 68).

The above mentioned traditions and legends are just some examples of the rich cultural heritage structured around the Nile in Egypt and Ethiopia. Identities structured around the Nile and the control of water have existed since time immemorial, and these identities have both been social and religious in nature pre-dating the nation-state, but today's identities build on the long historic trajectories, although in different ways. Dams have been seen as the useful pyramids (Schnitter, 1994), although the control of the Nile was of utmost importance for the ancient Egyptian pharaohs and the building of the pyramids can also be seen in water perspectives (Oestigaard, 2011). Still, dams have been and still are important in the creation of national identities within the framework of progress and modernity.

Control and security in the present

I will now move from the distant past and mythologies to the recent history starting with the 1959 Agreement. Relations between Egypt and Sudan, and also Ethiopia, have partly been structured around water, and national ideologies have not only shaped identities but also acknowledged water as a resource to be fought over. Sudan gained its independence on 1 January 1956. She subsequently declared a unilateral non-adherence policy towards the 1929 Agreement, and was particularly concerned about the Aswan High Dam. Thus, with independence Sudan demanded more of the waters of the Nile, thereby jeopardizing Egypt's plan to construct the Aswan High Dam. In response, Egypt mobilized troops at the Sudanese border (Arsano, 2007: 88). Tensions between Sudan and Egypt rose before the military takeover in Sudan in 1958, which enabled further negotiations over the utilization of the Nile. On 8 November 1959, the Agreement for the Full Utilization of the Nile Waters was signed in Cairo between Egypt and Sudan, but not including the other riparian states (Waterbury, 1979: 70–72). Then, as now, it is argued from an Egyptian perspective:

> The Nile River is Egypt's principle artery of life. It is life itself for Egypt. This basic fact does not apply to the same extent to the other riparian states. Therefore, one of the major strategic threats to Egyptian national security is the threat to vital resources lying beyond Egyptian borders.
>
> (Hassan and Rasheedy, 2011: 131)

When the Aswan High Dam in Egypt was inaugurated in 1971, it was one of the biggest dams in the world. The length at the crest of the Aswan High Dam is 3,600 metres and the height above bed level 111 metres. The total capacity of the reservoir is 162bcm (billion cubic metres), and its length is 500 kilometres, with a surface area of around 5,000 square kilometres (Waterbury, 1979: 111).

The importance of the Aswan High Dam was described by Nasser in this way in 1958:

> For thousands of years the Great Pyramids of Egypt were foremost among the engineering marvels of the world. They ensured life after death to the Pharaohs. Tomorrow, the gigantic High Dam, more significant and seventeen times greater than the Pyramids, will provide a higher standard of living for all Egyptians.
>
> (Joesten, 1960: 59)

This was essential for his grand modernization project, and Nasser aimed to turn Egypt into the 'Japan of Africa' (Tvedt, 2010: 7). The dam represented Egyptian nationalism and independence from the past (Collins, 2008). As Gamal Hamdan has remarked: 'The High Dam has simply eliminated the flood … It has converted the destructive river into a giant irrigation canal' (Waterbury, 1979: 87). The dam was constructed to store the equivalent of two annual floods, so that if the floods in one year were disastrously low, Egypt would still have enough water until the next flood. The benefits of the Aswan High Dam were demonstrated in 1972 when the flood was one of the lowest in the century. Without the release of the waters behind the dam, over one-third of Egypt's harvest would have been lost (Benedick, 1979: 140). With the Aswan High Dam, the water security was controlled within Egypt's borders.

The Aswan High Dam (111 metres) was not only seen as an engineering achievement, but the 'height' was also a symbol of the nation-building process of that time. During construction, the dam hired at least 30,000 constructors who took national pride in the project, symbolizing also their own achievements and sacrifices, and they also referred to themselves as being soldiers of an army. As one worker recalled:

> By October of 1952 the Dam was being studied; only four months after the Revolution! … Once the state told the story, that we went to the World Bank and it wouldn't give us money and that this was a challenge people didn't want us to achieve, we were ready to do anything it takes. The Dam's story is the story of a people; the struggle of a people.

Another construction worker was even more explicit: 'The Dam was a revolution within the Revolution. If the military Revolution liberated the country, the Dam liberated the people. It helped us to liberate ourselves from ourselves. It liberated us from fear.' Deaths during the construction happened, but were still honourable: 'This Dam was built with the bodies of its martyrs (Mossallam, 2014: 300, 302, 307). Still, despite nationalism and revolutionary war rhetoric, 'Although the dam changed the Nile's behaviour in Egypt, it did not liberate the country and the political actors from the power of the structure of the water system itself. The dam made the Egyptians more dependent on the Nile than ever before' (Tvedt and Coopey, 2010: 16). Hence, from an Egyptian perspective:

> The Nile River will always be the defining influence of Egyptian foreign policy towards the states in the basin region. The issue of the Nile River water and its significance for the survival of the people in Egypt has been

> clear in the minds of all political leaders in Cairo to the extent that there has been no significant difference among the successive administrations over the decades.
>
> (Hassan and Rasheedy, 2011: 149)

In May 1978, Egyptian President Sadat warned: 'We depend upon the Nile 100 per cent in our life, so if anyone, at any moment thinks to deprive us of our life we shall never hesitate [to go to war] because it is a matter of life or death' (Waterbury, 1979: 78). In 1979, Sadat reiterated that 'the only matter that could take Egypt to war again is water', and Egypt has threatened Ethiopia with war on numerous occasions. In 1977, for instance, when Ethiopia planned to divert water from the Blue Nile for irrigation schemes in response to frequent droughts and famines, Sadat declared that if Ethiopia proceeded, 'there is no alternative to use our force. We will retaliate when something happens but we have to be ready with plans and alternatives to firmly stop any action' (Adar, 2011: 182–183). Moreover, although relations between Egypt and Sudan regarding the utilization of the Nile have generally been cooperative, Egypt has nevertheless on several occasions threatened military force against Sudan to protect Egypt's interests in the Nile's waters (Deng, 2011: 39). President Mubarak, for instance, once said: 'those who play with fire in Khartoum … will push us to confrontation and to defend our rights and our lives' (Deng, 2011: 51). In fact, Deng argues, 'Egypt has issued … more threats against Sudan than against any other country of the Nile Basin. Hence, the duality is more … an Egyptian imposition that the Sudanese regimes bitterly swallow, lest they become easy prey to Egyptian political manipulations and plots' (Deng, 2011: 52–53).

Just before and after the CFA (Cooperative Framework Agreement) was signed on 14 May 2010, the rhetoric adopted by Egyptian politicians was aggressive. Mohamed Nasreddin Allam, minister of irrigation and water resources, made a number of harsh statements in newspapers and to parliament in this period: 'The new treaty is not binding on us. It will only be an obligation for the countries that signed it. At the same time we will make sure that this treaty won't affect Egypt's share of the Nile's water.'[3] The vulnerability of Egypt was stressed: 'Ask the Egyptians to leave their culture and go and live in the desert because you need to take this water and to add it to other countries? No.'[4] He went on to emphasize Egypt's claim of historic rights to the Nile: 'Egypt's share of the Nile's water is a historic right that Egypt has defended throughout its history', and, 'We will not sign on to any agreement that does not clearly state and acknowledge our historical rights.' Finally, he stressed the national security aspect of water for Egypt: 'Nile water is a matter of national security to Egypt. We won't under any circumstances allow our water rights to be jeopardized.'[5] He continued: 'Egypt reserves the right to take whatever course it sees suitable to safeguard its share.'[6] Similarly, prior to the signing of the agreement in May, Moufid Shehab, Egyptian minister of legal and parliamentary affairs, said that 'Egypt's historic rights to Nile waters are a matter of life and death. We will not compromise them.'[7] Former Defence Minister Field Marshal Mohammed Hussein Tantawi also once remarked: 'Egypt's defence strategy was carved-out with a possible war over the Nile very much in our mind.'[8] In similar vein, one Member of Parliament asserted in April 2010 that 'we welcome this war if it is imposed on us.'[9]

Ethiopia also needs more Nile water for development and its 'consistent position has been that it must make use of its water resources in general and the Nile waters in particular to develop the impoverished country' (Arsano, 2010: 174). In the 1950s, Haile Selassie stated:

> [I]t is of paramount importance to Ethiopia, a problem of the first order that the waters of the Nile be made to serve the life and the needs of our beloved people now living and those who will follow us in centuries to come. However, generally, Ethiopia may be prepared to share this tremendous God given wealth of hers with friendly nations neighbouring upon her, for the life and welfare of their people, it is Ethiopia's sacred duty to develop the great watershed which she possesses in the interests of her own rapidly expanding population and economy. To fulfil this task, we have arranged for the problem to be studied in all its aspects by experts in the field. Ethiopia has time and again set this forth as her position regarding the utilization of the Nile waters.
>
> (Op. cit. Arsano 2010: 165–166)

Controlling water for the benefit of a country's inhabitants is fundamental for a nation state and hence part of a national ideology whether in Egypt, Ethiopia or elsewhere. From one perspective, as a national ideology citizens do not need to share all parts of the ideology, but the outcome – health and wealth in a broad sense – concerns everyone. Control of water by dams – and irrigated agriculture – is seen as fundamental for Ethiopia's development and food security. In 1981 Amartya Sen, who was awarded the Nobel Prize in Economic Sciences in 1998, published his acclaimed *Poverty and Famines: An Essay on Entitlement and Deprivation*. While discussing the 1971–1973 drought in Ethiopia, this was largely due to the failure of the rains, but as he says,

> since the Ethiopian famine clearly was initiated by a drought, and since drought causes crop failures (and, indeed, did so in this case), it is easy to be predisposed towards accepting an explanation of the famine in terms of food available. But a drought causing an agricultural or pastoral crisis not only reduces food supply; it also cuts the earnings of the agriculturist or the pastoralist, affecting his command over food.
>
> (Sen, 1981: 88–90)

A fundamental duty of the state is to provide wealth and security for her inhabitants, and the Ethiopian drought in 1984–1985 clearly showed the importance of dams since Ethiopians, living in 'Africa's water tower', died of drought, while Egypt could prosper or continue life because of the stored water in the Aswan Dam, as noted above during the 1971–1973 droughts and also during the 1984–1985 droughts.

In a notable twist of history of Nile geopolitics, on 30 March 2011, just one and a half months after the fall of the Mubarak regime, Ethiopia announced its biggest ever project on the Blue Nile, the Grand Ethiopian Renaissance Dam (GERD) at the border with Sudan. This dam will be for hydropower and not irrigated agriculture. When it is completed in 2017, it will be Africa's largest and the world's tenth largest dam with a reservoir of about 70bcm and an estimated cost of about US$4.7 billion. As part of the construction work, for the first time in history the Nile's flow was diverted in 2013: 'The dam is being built in the middle of the river so you can't carry out construction work while the river flowed', Mihret Debebe said. 'This now enables us to carry out civil engineering work without difficulties. The aim is to divert the river by a few metres and then allow it to flow on its natural course.'[10] Egypt's Deputy Foreign Minister for African Affairs, Ali Hifni, said that the diversion of the river was not something to worry about, but that the dam itself was a concern.[11] A week later, however, a meeting called by the Egyptian President Mohammed Morsi was broadcasted live on TV, apparently by mistake without the consent of the politicians, where they openly discussed various military actions towards Ethiopia. 'One of the politicians suggested sending special forces to destroy the dam; another thought of jet fighters to scare the Ethiopians; and a third called for Egypt to support rebel groups fighting the government in Addis Ababa.'[12] Although not the first time such threats have been put forward, what makes this unique is perhaps the context where Ethiopia for the first time in history and mythologies apart had actually started to divert the river, evoking a 1,000-year-old legend and religious beliefs about controlling the Nile.

Although Ethiopia has constructed and is currently constructing several large dams (see Figure 11.3), the GERD has a unique place in creating a national ideology and identity around the Nile and control of water. The way the late President Meles Zenawi played the Nile card, externally and internally, was masterly done, seen from the Ethiopian perspective. Even stern critics and dissidents of the regime support the dam and the financial strategy for funding the construction by buying bonds. Not only does the dam portray Ethiopian greatness and control over water in relation to Egypt, but when the World Bank, European Union and the African Development Bank did not fund the dam, not even the Chinese, who fund and build dams everywhere, the Ethiopians did it by themselves (although there are uncertainties regarding the funding). This created a strong sense of identity among many Ethiopians: 'We can do it!' or 'We did it!', as I was told by many in 2011, despite obstacles from the rest of the world. As pointed out by one in 2015, although with a twist of humour, the GERD shows to Egypt, the World Bank and the world, in that chronological order, that Ethiopia can build large dams, and there is undoubtedly such perceptions of pride and possibilities involved in the national project. Moreover, throughout the country in different towns and villages, the construction of the dam is celebrated by huge processions where trophies showing victory are displayed.

Donation of money to the Nile is, however, nothing new, a churchman at Gish Abay pointed out in 2015, which resonates well with Ethiopian traditions of offering small sacrifices to the river. Although in the current political context this is done to construct the dam, Ethiopians have always donated offerings to their rivers and water, like bread or other offerings, while, for instance, crossing a river. Thus, people's donations to the dam builds on and continues a long tradition of giving and making offerings to the Nile, he said, which may also explain the deep cultural resonance the dam evokes.

Figure 11.3 Ribb Dam in the Lake Tana region under construction

Source: Terje Oestigaard.

Thus, it is without doubt that control over water may prompt strong national identities if the politics are orchestrated along these lines – in Egypt and Ethiopia as elsewhere. Large dams controlling the waters are often fundamental in the national identity processes, and herein seemingly is a paradox. While the large water structures such as dams may contribute to the creation of an overall national identity, they often conceal domestic conflicts. Dams do displace people and usually face a lot of opposition internally and externally, but as Wondwosen points out in Chapter 9, those enabling and partaking in large-scale land acquisitions can be seen as 'development heroes' by governmental officials rather than doing harm to the country and minority groups.

Conclusion

It is difficult to make any conclusive statements about changing identities and the role of national ideologies since they are, like the Nile itself, fluid. As a preliminary hypothesis it seems, however, that if a nation has been successful in making a national ideology or identity structured around controlling water because of scarcity or absolute needs, it also gives a greater flexibility and leverage in implementing development projects that, at local and individual levels, may even be perceived as jeopardizing and contradicting these very national aims, like building dams. At the same time, strong national Nile identities may prevail despite suffering and hunger of people because of overall national politics. Long historic trajectories and continuities in water mythologies and histories may strengthen these individual identities. Within the frame of modernizing projects and processes, autonomous states are expected to act favourably, internally and externally. From a national perspective, proactive development and control over the Nile resources may strengthen unity and identities; absence of ambitions and plans may weaken the national project and be more open to criticism from within and beyond. In short, a successful national ideology is one of the strongest tools ever possible to imagine or use, and, if water is a key component in such identities, this should preferably be included in analyses of changing hydropolitics.

Notes

1 See www.oxforddictionaries.com/definition/english/ideology (accessed 26 January 2016).

2 The following ethnographic description is based on my own fieldworks and published in Oestigaard and Gedef (2013).

3 *Al-Ahram Weekly*, 1–7 July 2010, http://weekly.ahram.org.eg/2010/1005/eg4.htm (accessed 9 March 2012).

4 *Aljazeera*, 28 June 2010, www.aljazeera.com/news/middleeast/2010/06/201062871134786105.html (accessed 9 March 2012).

5 *Los Angeles Times World*, 20 April 2010, http://latimesblogs.latimes.com/babylonbeyond/2010/04/egypt-minister-rejects-nile-sharing-deal-as-experts-warn-of-water-famine.html (accessed 9 March 2012).

6 *Sudan Tribune*, 11 May 2010, www.sudantribune.com/Sudan-formally-rejects-the-Nile,35041 (accessed 9 March 2012).
7 Government of Southern Sudan Natural Resource Management Group, 29 April 2010, www.ssudan-nrmg.org/nrmg/index.php/nrnewssouthsudan/1-latest-news/216-wateranalysis-egypt-spat-fuels-water-tension-in-nile-basin (accessed 9 March 2012).
8 *Tigrai State*, 2 July 2010, http://tigraionline.blogspot.com/2010_07_01_archive.html (accessed 9 March 2012).
9 *Guardian*, 26 April 2010, www.guardian.co.uk/commentisfree/2010/apr/26/egypt-nile-water-negotiations (accessed 9 March 2012).
10 Ethiopia diverts Blue Nile for controversial dam build. BBC, 28 May 2013, www.bbc.co.uk/news/world-africa-22696623 (accessed 9 September 2013).
11 Ibid.
12 Egyptian politicians caught in on-air Ethiopia dam gaffe. BBC, 4 June 2013, www.bbc.co.uk/news/world-africa-22771563 (accessed 9 September 2013).

References

Adar, K. G. (2011). Kenya's foreign–policy and geopolitical interests: The case of the Nile River Basin. In Adar, K. G. and Check, N. A. (eds). *Cooperative Diplomacy, Regional Stability and National Interests: The Nile River and the Riparian States*: 167–188. Pretoria: Africa Institute of South Africa.

Anderson, B. (1993). *Imagined Communities*. London: Verso.

Arsano, Y. (2007). *Ethiopia and the Nile. Dilemmas of National and Regional Hydropolitics*. Zurich: ETH.

Arsano, Y. (2010). Institutional development and water management in the Ethiopian Nile Basin. In Tvedt, T. (ed.) *The River Nile in the Post-Colonial Age*: 161–178. London: I.B. Tauris.

Ascherson, N. (1996). Foreword. In Atkinson, J. A., Banks, I. and O'Sullivan, J. (eds). pp. v–ix. *Nationalism and Archaeology*. Glasgow: Scottish Archaeological Forum.

Bell, D. N. (transl.) (1983). *Besa. The Life of Shenoute. Introduction, Translation, and Notes*. Michigan: Cistercian Publications.

Benedick, R. E. (1979). The High Dam and the transformation of the Nile. *Middle East Journal*, 33 (2): 119–144.

Brown, K. S. (1998). Contests of heritage and politics of preservation in the Former Yugoslav Republic of Macedonia. In Meskell, L. (ed.) *Archaeology under Fire: Nationalism, Politics and Heritage in the Eastern Mediterranean and Middle East*: 68–86. London: Routledge.

Collins, R. (2008). Managing the water of the Nile: Basis for cooperation? In Gershoni, I. and Hatina, M. (eds). *Narrating the Nile: Politics, Cultures, Identities*: 181–201. Boulder and London: Lynne Rienner Publishers.

Deng, B. K. (2011). Cooperation between Egypt and Sudan over the Nile River waters: The challenges of duality. In Adar, K. G. and Check, N. A. (eds). *Cooperative Diplomacy, Regional Stability and National Interests: The Nile River and the Riparian States*: 39–66. Pretoria: Africa Institute of South Africa.

Erlich, H. (2002). *The Cross and the River. Ethiopia, Egypt and the Nile*. London and Boulder: Lynne Rienner Publishers.

Gellner, E. (1997). *Nationalism*. New York: New York University Press.

Hassan, H. A. and Rasheedy, A. A. (2011). The Nile River and Egyptian foreign-policy interests. In Adar, K. G. and Check, N. A. (eds). *Cooperative Diplomacy, Regional Stability and National Interests: The Nile River and the Riparian States*: 131–152. Pretoria: Africa Institute of South Africa.

Himdan, J. (1987). *The Character of Egypt*. Cairo: Alam Al Kotob.
Insoll, T. (2004). Are archaeologists afraid of gods? Some thoughts on archaeology and religion. In Insoll, T. (ed.). *Belief in the Past: The Proceedings of the Manchester Conference on Archaeology and Religion*: 1–6. BAR International Series 1212, Oxford.
Jackson, R. (2007). *Sovereignty*. Cambridge: Polity Press.
Jenkins, R. (2004). *Social Identities*. London: Routledge.
Joesten, J. (1960). Nasser's daring dream: The Aswan High Dam. *World Today*, 16 (2): 55–63.
Kampanje-Phiri, D. (2015). Nationalism, state sovereignty and international water management in the Zambezi River. In Tvedt, T., McIntyre, O. and Kassa Woldetsadik, T. (eds). *A History of Water, Series 3, Volume 2: Sovereignty and International Water Law*: 578–600. London: I.B. Tauris.
Krawiec, R. (2002). *Shenoute and the Women of the White Monastery: Egyptian Monasticism in Late Antiquity*. Oxford: Oxford University Press.
Lutfi, H. (1998). Coptic festivals of the Nile: Aberrations of the past? In Philipp, T. and Haarmann, U. (eds). *The Mamluks in Egyptian Politics and Society*: 254–282. Cambridge: Cambridge University Press.
Mossallam, A. (2014). We are the ones who made this dam 'High'! A builders' history of the Aswan High Dam. *Water History*, 6: 297–314.
Oestigaard, T. (2007). *Political Archaeology and Holy Nationalism: Archaeological Battles over the Bible and Land in Israel and Palestine from 1967–2000*. Gothenburg: Gotark. C No. 67.
Oestigaard, T. (2009). Christianity and Islam as Nile religions in Egypt: Syncretism and continuity. In Oestigaard, T. (ed.). *Water, Culture and Identity: Comparing Past and Present Traditions in the Nile Basin Region*: 141–164. Bergen: BRIC Press.
Oestigaard, T. (2011). *Horus' Eye and Osiris' Efflux: The Egyptian Civilisation of Inundation ca. 3000–2000 BCE*. Oxford: Archaeopress.
Oestigaard, T. and Gedef, A. F. (2013). *The Source of the Blue Nile: Water Rituals and Traditions in the Lake Tana Region*. Newcastle: Cambridge Scholars Press.
Pankhurst, R. (2000). Ethiopia's alleged control of the Nile. In Erlich, H. and Gershoni, I. (eds). *The Nile: Histories, Cultures, Myths*: 25–38. London: Lynne Rienner Publishers.
Ranger, T. (1973). Territorial cults in the history of Central Africa. Journal *of African History*, 14 (4): 581–597.
Sanders, T. (2002). Reflections on two sticks: Gender, sexuality and rainmaking. *Cahiers d'Études Africaines*, 42, Cahier 166: 285–313.
Schnitter, N. J. (1994). *A History of Dams: The Useful Pyramids*. Rotterdam: A. A. Balkema.
Sen, A. (1981). *Poverty and Famines: An Essay on Entitlement and Deprivation*. Oxford: Clarendon Press.
Simonse, S. (1992). *Kings of Disaster: Dualism, Centralism and the Scapegoat King in Southeastern Sudan*. Leiden: E. J. Brill.
Six, V. (1999). Water, the Nile, and the Tä'amrä Maryam: Miracles of Virgin Mary in the Ethiopian Version. *Aethiopica*, 2: 53–68.
Skarstein, R. (2016). Primitive accumulation: Concept, similarities and varieties. In Havnevik, K., Oestigaard, T., Virtanen, T. and Tobisson, E. (eds). *Framing African Development. Challenging Concepts*: 135–168. Leiden: Brill.
Sørensen, M. L. S. (1996). The fall of a nation, the birth of a subject: The national use of archaeology in the nineteenth-century Denmark. In Díaz-Andreu, M. and Champion, T. (eds) *Nationalism and archaeology in Europe*: 24–47. London: UCL Press London.

Tvedt, T. (2010). About the importance of studying the modern history of the countries of the Nile Basin in a Nile perspective. In Tvedt, T. (ed.) *The River Nile in the Post-Colonial Age*: 1–11. London: I.B. Tauris.

Tvedt, T. and Coopey, R. (2010). A 'water systems' perspective on history. In Tvedt, T, and Coopey, R. (eds). *A History of Water, Series 2, Volume 2: From Early Civilizations to Moderen Times*: 1–26. London: I.B. Tauris.

Tvedt, T. and Oestigaard, T. (2010). A history of the ideas of water: Deconstructing nature and constructing society. In Tvedt, T. and Oestigaard, T. (eds). *A History of Water. Series 2, Volume 1. The Ideas of Water from Antiquity to Modern Times*: 1–36. London: I.B. Tauris.

Tvedt, T., McIntyre, O. and Kassa Woldetsadik, T. (2015). Sovereignty, the web of water and the myth of Westphalia. In Tvedt, T., McIntyre, O. and Kassa Woldetsadik, T. (eds). *A History of Water, Series 3, Volume 2: Sovereignty and International Water Law*: 3–26. London: I.B. Tauris.

Waterbury, J. (1979). *Hydropolitics of the Nile Valley*. New York: Syracuse University Press.

Wild, R. A. (1981). *Water in the Cultic Worship of Isis and Sarapis*. Leiden: E. J. Brill.

Young, D. W. (1993). *Coptic Manuscripts from the White Monastery: Works of Shenute*. Vienna: Hollinek.

Index

For Product Safety Concerns and Information please contact our EU
representative GPSR@taylorandfrancis.com
Taylor & Francis Verlag GmbH, Kaufingerstraße 24, 80331 München, Germany

www.ingramcontent.com/pod-product-compliance
Ingram Content Group UK Ltd.
Pitfield, Milton Keynes, MK11 3LW, UK
UKHW022134190926
475702UK00013B/1798
* 9 7 8 0 3 6 7 0 2 9 6 9 2 *